STUDY GUIDE AND SELECTED SOLUTIONS MANUAL

Julie Frentrup • Suzanne Saum

Fundamentals of Chemistry

Fourth Edition

Ralph A. Burns

Upper Saddle River, NJ 07458

Project Manager: Kristen Kaiser
Senior Editor: Kent Porter-Hamann
Editor in Chief: John Challice
Executive Managing Editor: Kathleen Schiaparelli
Assistant Managing Editor: Dinah Thong
Production Editor: Natasha Wolfe
Manufacturing Manager: Trudy Pisciotti
Supplement Cover Management/Design: Paul Gourhan
Manufacturing Buyer: Ilene Kahn

Pearson Education, Inc.
Upper Saddle River, NJ 07458

The author and publisher of this book have used their best efforts in preparing this book. These efforts include the development, research, and testing of the theories and programs to determine their effectiveness. The author and publisher make no warranty of any kind, expressed or implied, with regard to these programs or the documentation contained in this book. The author and publisher shall not be liable in any event for incidental or consequential damages in connection with, or arising out of, the furnishing, performance, or use of these programs.

Printed in the United States of America

10 9 8 7 6 5 4 3 2 1

ISBN 0-13-033779-X

Pearson Education Ltd., *London*
Pearson Education Australia Pty. Ltd., *Sydney*
Pearson Education Singapore, Pte. Ltd.
Pearson Education North Asia Ltd., *Hong Kong*
Pearson Education Canada, Inc., *Toronto*
Pearson Educacíon de Mexico, S.A. de C.V.
Pearson Education—Japan, *Tokyo*
Pearson Education Malaysia, Pte. Ltd.
Pearson Education, *Upper Saddle River, New Jersey*

TABLE OF CONTENTS

Chapter

1	Chemistry is Everywhere	1
2	Matter and Energy	3
3	Fundamental Measurements	11
4	Elements, Atoms, and the Periodic Table	19
5	Atomic Structure: Atoms and Ions	25
6	Names, Formulas, and Users of Inorganic Compounds	31
7	Periodic Properties of Elements	37
8	Chemical Bonds	43
9	Chemical Quantities	53
10	Chemical Reactions	61
11	Stoichiometry: Calculations Based on Chemical Equations	73
12	Gases	81
13	Liquids and Solids	91
14	Solutions	99
15	Reactions Rates and Chemical Equilibrium	107
16	Acids and Bases	115
17	Oxidation and Reduction	125
18	Fundamentals of Nuclear Chemistry	133
19	Organic Chemistry	143
20	Biochemistry	151

CHAPTER 1

Chemistry Is Everywhere

During this course, THINK about what you are reading and hearing. Do not merely memorize definitions and procedures. Think about the *concepts* presented to you. Then you will develop an understanding of chemistry.

Try to keep an open mind. Too often the media present a very negative view of chemistry. Chemicals themselves are *value free* materials to which people attach positive or negative values, often without good reasons.

SKILLS TO ENHANCE SUCCESS IN THIS CHAPTER

Learning involves the acquisition of knowledge and the acquisition of skills. You acquire factual knowledge by rote memorization. You acquire skills by searching out the connections between the bits of your acquired knowledge, relating this knowledge to your daily life, and by practicing.

During your study of the first few chapters, you will be asked to solve both quantitative and qualitative problems. To solve quantitative problems you must do more than "plug-and-chug" through a formula to see if your answer matches the one in the book. **Carefully** read the applicable text section and apply its concepts to each problem. Look for the "big picture" by comparing the problems and the underlying concepts they have in common; there is a pattern.

You will not immediately perfect any skill, but will make steady progress by repetition. Think of examples from your everyday life and relate them to the concepts presented. Often study groups will help deepen your understanding. Rewrite definitions of concepts in your own words and explain them to others. Assemble a personal strategy for solving problems. As the course progresses, remember that you need all prior skills to succeed. Keep reviewing these skills. Chemistry is a continuous study, which builds upon the foundation laid in earlier chapters. The least effective way to study is cramming at the last minute before an exam. Avoid this under all circumstances.

CHAPTER 1 PROBLEMS

1.1a Which of these is NOT a chemical?
A) aspirin B) sugar C) oil D) light E) water
Answer: D Light is not a chemical. (It is energy.)

1.1b Name a material that is NOT made of chemicals. **Answers to part "b" are collected after the Problems.**

1.2a Explain why "gravity" is called a law.
Answer: Gravity is supported by consistent experimental results and has no known exceptions.

1.2b Identify this statement as law, theory, or hypothesis: "If I fail to water my plant for three days, it will die."

1.3a Would a quality-control chemist, who tests products for quality during production, be involved in basic research or applied research? Explain your answer.
Answer: This is applied research because of his/her involvement with a marketable product.

1.3b Give an example of applied chemical research.

1.4a Why is it necessary and/or useful for an art major to be exposed to chemistry?
Answer: Many art supplies are mixtures that contain metals heavy metals such as lead and mercury, and other toxic materials. Some ceramic glazes can be very dangerous to use.

1.4b Do you need a knowledge of chemistry to be an informed citizen?

Answers to "b" Problems

1.1b This cannot be done, since all "material" is matter. Matter is made up of chemicals. Energy is not a material, so it would not qualify.

1.2b Hypothesis. The statement is reasonable, but this may or may not happen depending upon conditions such as the health of the plant, its moisture requirements, and the temperature.

1.3b Applied research would include a study designed to make a paint pigment more resistant to fading, or experiments to determine a less expensive way to synthesize a particular medicine. Any example that involves specific investigation of an existing product to improve its marketabilty, safety, profitability, or performance would be applied research.

1.4b We hope you are convinced that the answer is yes.

PRACTICE TEST I

1. Which of the following chemicals is most likely to have a negative use?
   A) ibuprofen B) helium C) methamphetamine D) water E) mercury
2. Which of the following inventions required a chemist during its development?
   A) computer chip B) automobile paint C) pesticide D) all of these
3. In order to find the number of inches in seven feet, what known fact must you provide?
4. To find out how many square feet of carpet to buy for a room in your home, what data must you collect?
5. The "scientific method" is one example of what cycle?
6. Give at least two characteristics that differentiate an experiment from any other task.
7. Which of the following can be proved by experimentation?
   A) natural law B) hypothesis C) theory D) all of these
8. Explain the difference between basic and applied research.
9. Give at least one reason why an attorney should have a knowledge of chemistry.

Matching: Describe the following terms.

10. a natural law — A) an established explanation that has been tested over time
11. a hypothesis — B) a proposed reasonable explanation that has not yet been proved
12. a theory — C) a summary of known facts not yet refuted or proved false

Answers to Practice Test I

1. C
2. D
3. You need the fact that there are twelve inches in one foot.
4. You need the length and width of the room, which, when multiplied, give the floor area.
5. The scientific method is an example of the steps used in problem solving.
6. Experiments are controlled, explain the composition and actions of nature, and are repeated many times.
7. D
8. The use of the results differentiates applied from basic research. Basic research is acquiring knowledge for its own sake. Applied research relates to the production of a marketable product.
9. Many issues before our courts involve science. Legislators are passing more laws that regulate scientific and technical activity. An attorney with a knowledge of chemistry can more intelligently deal with these cases.

10. C 11. B 12. A

CHAPTER 2

Matter and Energy

Do you sometimes say, "I don't have any energy today."? What do you mean? Can you express exactly how much energy you "have"?

All sciences study *matter* and *energy* in one form or another. To our senses, matter and energy have entirely separate existences. This attitude was taken in "classical" science. Then Einstein showed that matter and energy are two aspects of the same underlying phenomenon. Since we do not yet understand this underlayment, we commonly explain Einstein's conclusion by saying that matter and energy are interchangeable.

Chemistry continues to treat matter and energy as distinct realities. Chemistry studies the interaction of substances in what are called chemical reactions. That a reaction has occurred is evidenced by formation of new substances and an interchange of energy. During a chemical reaction, heat energy is given off or absorbed. Most common reactions we are aware of liberate (give off) energy. These reactions heat our homes, cook our food, and propel our automobiles. Chemical reactions also supply the energy needed to produce our electricity.

SKILLS TO ENHANCE SUCCESS IN THIS CHAPTER

In Chapter 2 we introduce these skills:

- Observing the world about us.
- Making careful distinctions between our observations.
- Expressing these distinctions to others in communication.
- Making conclusions from our observations about a world hidden from our senses.

Throughout our lives we learn - as we did when we were children - by asking questions. Let's look at a few questions applicable to the concepts of this chapter.

What is mass? We speak of a massive object, a massive weather pattern, a massive injury. Common use of the term connotes large quantity or size. But *mass* has specific meaning in science. It may be large or small. It is not the same as *weight*.

What is an element? We commonly speak of the elements of nutrition, the elements of music, the elements of any study. Initial courses of study are termed elementary.

What is a mixture? A recipe might direct you to mix ingredients and then refer to the resulting combination as a mixture. What is a solution? You put sugar in your coffee and stir to mix. You obtain a solution of sugar and coffee in water. Both examples are mixtures, but only the second one is a solution.

What is a substance? A compound? A material? Why does the law use the term "controlled substance" rather than "controlled material"? A controlled "substance" may actually be a *mixture*, not a substance, chemically speaking - for example, cocaine cut with sugar. Chemists distinguish two substances in this mixture; the law doesn't pay attention to the difference.

We say that a mixture is "composed" of certain substances. We form compounds by reacting elements together and say that a compound is "composed" of certain elements. But the two "composeds" are different.

Many of the above terms are everyday words and phrases, but in science these terms have special, stricter meanings. The scientific definitions of these words may be similar to the everyday meaning but are narrower than the nonchemical word. This pickiness keeps our thinking clear and improves the communication of thoughts and ideas with others. It is necessary, not just a vanity or conceit, to be more particular about how we use these words.

When learning a new subject, classifying the various observed aspects of the subject greatly aids our thinking. We classify our world into matter and energy. We further classify matter into substances and mixtures, substances down into elements and compounds, and so on.

Observe the world about you — your home, your workplace, your environment. Chemistry is not limited to the classroom or laboratory. Far from it. Chemistry Is Everywhere, as Chapter 1 proclaims.

CHAPTER 2 PROBLEMS

We present problems in pairs, designated a and b, for each text section for which problems are applicable. We give a complete solution to part a You are expected to solve part b yourself. Answers to "b" problems are given after the problems.

We sometimes present a problem solution method differing from the method of the text for the same type of problem. At times we present more than one method. Study all methods and develop a solution method that works for you. Each of us eventually develops his/her own method of solving problems, based on methods that we have been taught, but they may differ in the details.

2.1a. You take a thrill ride on one of the roller coasters that goes through a complete vertical loop. At the top of the loop you are upside down. What is your weight at the top of the loop?
Answer: Since you are upside down and not falling out of the car, your weight must be close to zero. You have a seat belt on or a bar across your lap, but you do not experience a feeling of falling out of the seat. Centrifugal force is counteracting gravity at the top of the loop. You are temporarily weightless.

2.1b. Are our astronauts, when in orbit around the earth, in a weightless environment? Explain your answer.

2.2a. During boiling a substance changes from a liquid state to a gaseous state. Can a liquid become a gas at a temperature below its boiling point? How would you know this without being told?
Answer: Yes, a liquid will vaporize below its boiling point. For example, water will evaporate at room temperature. Also, certain liquids have an odor. But we can smell only gases. The liquid must have vaporized into a gas, which enters the nose and activates olfactory sense organs and nerves. If a mist (liquid) or smoke (solid) enters our nostrils, we may sneeze or cough from it, but we will not smell the solid or liquid part, only the gases that emanate from them.

2.2b. Can a substance exist in two or more states at the same time?

2.3a. Metals are obtained from their ores either by chemical reactions such as smelting (reacting with coke at high temperatures), or by electrolyzing (passing an electric current through a melt of the refined ore). Is the ore a mixture or a compound? What really is an ore?
Answer: Raw ores from the earth's crust are actually mixtures that usually contain much ordinary rock and sand, mixed in during the mining process. The raw material is refined by crushing and removing the nonusable parts. In a chemical sense, this refined material is the ore. This refined ore, from which the metal is obtained, is a compound.

2.3b. Is aluminum an element or a compound?

2.4a. Is an aspirin tablet a pure substance or a mixture?
Answer: Read the label on the bottle of aspirin tablets. The tablet contains various binders, so-called pelletizers, to facilitate forming the tablet and holding it together until use, and extenders to make the tablet a practical size. These are cellulose materials, starches, glycols, silica, and acids. The aspirin tablet is a mixture.

2.4b. Is the aspirin itself a pure substance or a mixture?

2.5a. The cola in a bottle before opening is a homogeneous mixture, that is, a solution. What is in the bottle just after opening?
Answer: It is a heterogeneous mixture, since the gas bubbles are rising up through the liquid beverage.

2.5b. Name two pure substances found in a cola soft drink.

2.6a. When water "boils away" it seems to disappear. How could you indicate (not prove) the conservation of mass during boiling of water?
Answer: Hold a glass in the steam coming from the water. Droplets of water, condensate, form on the glass. This indicates that the water is not being destroyed. It is reasonable to assume that the process continues for the entire mass of the water. Or you could simply point out the condensation of water on the windows of your house during cold weather. You have probably noted that during cooking and showering, the windows (and perhaps the walls) of your house "sweat".

2.6b. How would you prove the conservation of mass during boiling of water?

2.7a. When you pay your light bill, what are you actually paying for? What is electricity?
Answer: Electricity is a form of energy. We use it to produce light and heat, and to run motors to do our work for us. In so doing, we are converting one form of energy into another. The division of energy into many forms is practical for engineering and commerce purposes. But each of these various forms is either potential or kinetic. Potential and kinetic are broad, basic classifications of energy.

2.7b. List at least five forms of energy and classify each as fundamentally potential or kinetic.

2.8a. What is the energy source of the sun? What is the energy source of our nuclear power plants?
Answer: In both cases, the energy source is atomic, or more properly, nuclear. The sun's reaction is primarily that of converting hydrogen to helium. Hydrogen is a "lighter" element than helium, so the reaction is called *fusion*, because two lighter atoms are converted to one heavier one. Our atomic power plants rely upon *fission*, converting heavier elements into lighter ones. We have been trying to control fusion reactions for almost 50 years, since the energy generation is much greater than that of fission reactions. But the hydrogen must be contained at extremely high pressure and temperature that have not yet been achieved.

2.8b. What reason(s) do we have for neglecting the conversion of matter into energy, or vice versa, during our chemical reactions?

Answers to "b" Problems

2.1b Yes, the astronauts are weightless. Otherwise they wouldn't float around in their spacecraft.

2.2b Yes. Water, for example, exists as a solid and liquid at 0°C, its freezing point at normal conditions. It exists as a liquid and gas at 100°C, its boiling point at normal conditions. Water can even exist as a solid, liquid, and gas all at once, under special high-pressure conditions called the triple point.

2.3b Aluminum is an element. It is obtained from an oxide ore by electrolysis.

2.4b The aspirin must be a substance. It is synthetically made, not extracted from a naturally occurring source. If the aspirin were itself a mixture, our labeling laws would require that the ingredients in the aspirin be listed separately.

2.5b Wood, like coal, contains mostly carbon. Elemental carbon is black. (This is what makes coal black.) But wood is usually not black (although ebony is quite dark). So the carbon in wood is not elemental carbon. It is mostly cellulose, a compound. The process used to make charcoal is called "destructive distillation". Wood's cellulose is decomposed to charcoal (carbon), turpentine, and other organic compounds. This process was at one time an important industry, as charcoal and turpentine were common in ordinary households.

2.6b Water, sugar (sucrose), phosphoric acid, and carbon dioxide are all chemicals found in colas. Read the ingredients on the can. Biological ingredients like spices, extracts, and oils are usually complex mixtures.

2.7b Electrical: potential. Energy manifests as voltage, which is called electrical potential.
Chemical: potential. The energy is there by virtue of chemical composition.
Atomic: potential. The energy is there by virtue of nuclear composition.
Light or heat: potential. There is no mass in motion.
Mechanical: kinetic. The push of pistons or the turning of screws or gears is a motion.
Magnetic: potential. The energy is there by virtue of internal arrangement of the atoms.
A weight above your head: potential. This is the classical potential energy of a mass raised to a height in a gravitational field. The weight falling: obviously kinetic. As a weight falls, it can run clocks or machines, or turn cogs and gears.
Hydraulic (a waterfall): kinetic. We generate electricity from the energy of water falling through turbines that drive generators. Water behind a dam: potential energy.
Gas at high pressure: potential. This is stored energy, not yet in motion. Pressurized gases drive our automobiles by pushing the pistons in the engine.

2.8b. The change in mass is so small as to be undetectable. The energy changes are very slight compared to those obtained during atomic reactions. When you study Chapter 18, try applying $E = mc^2$ to the energy changes of chemical reactions. The change of mass is far less than our measurement abilities permit us to detect.

PRACTICE TEST I

1. A lunar space module has a mass of 12,000 kilograms. The moon's gravity is one-sixth that of Earth. What would the module's mass be on the moon?

2. Which state of matter is composed of tightly packed particles arranged in an orderly array?

3. Air is

   A) a heterogeneous mixture of several compounds.

   B) a pure substance (if unpolluted).

   C) a gaseous compound.

   D) an alloy.

   E) a homogeneous mixture of several gases.

4. Ethyl alcohol has a melting point of –117°C and a boiling point of 78.3°C. Identify the state of matter of ethyl alcohol at 25°C.

5. Which is a chemical change?

   A) fruit rotting
   B) candle wax melting
   C) water boiling
   D) grinding coffee
   E) dew condensing on the grass

6. Which describes a physical property of a substance?

   A) It is flammable.
   B) It is reflective.
   C) It reacts with oxygen.
   D) It is explosive.
   E) It is corrosive.

7. Antoine Lavoisier demonstrated the law of conservation of matter by

   A) discovering that no mass was lost when mercuric oxide was heated.

   B) showing that energy could be stored without loss.

   C) discovering relativity.

   D) causing a metal to lose mass by removing electrons.

   E) losing his head on the guillotine.

8. Which is a pure substance?

   A) air
   B) iodized table salt
   C) 100% natural honey
   D) oxygen
   E) flour

9. The freezing of water to make ice is

   A) an exothermic chemical change.

   B) an endothermic chemical change.

   C) an exothermic physical change.

   D) an endothermic physical change.

   E) an exergonic physical change.

10. Which would NOT be a good example to demonstrate potential energy?

   A) a stretched spring
   B) a pressurized can
   C) a dropped shoe on the floor
   D) a rat trap set to go off
   E) a diver on the high board

11. Which is an element?

   A) H_2O B) HI C) CO_2 D) NH_3 E) N_2

12. Which is a gas?

   A) ice B) steam C) oil D) salt E) pizza

PRACTICE TEST II

1. The moon's gravity is one-sixth that of Earth. What would an astronaut's 60-pound backpack weigh on the moon?

2. Any bartender can tell you that ethyl alcohol and water can be mixed in any proportion and will not separate out later. Therefore, these two liquids are said to be ___________ .

3. Which of the following is a heterogeneous mixture?

   A) salt water
   C) a copper pipe
   B) water, H_2O
   D) potato salad
   E) sucrose (pure cane sugar)

4. Bronze, a metal alloy used to make statues, is

   A) a homogeneous mixture.
   B) an element.
   C) a heterogeneous mixture.

5. Which of these is a chemical property?

   A) viscosity
   C) color
   B) toxicity
   D) compressibility
   E) melting point

6. When a log burns, is the matter destroyed? Explain your answer.

7. When a stretched rubber band is released,

   A) its kinetic energy increases.
   B) its kinetic energy decreases.
   C) its potential energy increases.
   D) its energy remains unaffected.
   E) None of the above is correct.

8. Which of the following is a physical change?

   A) a log burning
   C) a leaf changing color
   B) an ice cube melting
   D) your car door rusting
   E) "burning off" fat with exercise

9. Einstein's theory of relativity states that matter can be converted into energy, and in the process mass is not conserved. True or false?

10. Sodium chloride is

    A) a pure element.
    C) a homogeneous mixture.
    B) a mixture of two elements.
    D) a heterogeneous solution.
    E) a pure compound.

11. Inside your car's engine cylinders, a mixture of gasoline and air is ignited by a spark. The resulting process is

    A) an exothermic chemical change.
    B) an endothermic chemical change.
    C) an exothermic physical change.
    D) an endothermic physical change.
    E) an exergonic physical change.

12. Which has the most potential energy?

    A) a pebble rolling down a hill
    B) a boulder rolling down the same hill
    C) a pebble stopping at the bottom of the hill
    D) a boulder resting on top of the hill
    E) a boulder being pushed up the hill

Answers to Practice Test I

1. 12,000 kg (The mass of the module would not change. Only weight is affected by gravity.)
2. the solid state
3. E) a homogeneous mixture of several gases
4. Liquid (Solid below -117°C, liquid between -117°C and 78.3°C, gas above 78.3°C)
5. A) fruit rotting
6. B) It is reflective.
7. A) discovering that no mass was lost when mercuric oxide was heated.
8. D) oxygen (*Iodized* table salt has a small amount of sodium iodide added in to the sodium chloride.)
9. C) an exothermic physical change.
10. C) a dropped shoe on the floor
11. E) N_2
12. B) steam

Answers to Practice Test II

1. 10 pounds (one-sixth of the weight)
2. miscible
3. D) potato salad
4. A) a homogeneous mixture (Alloys are solutions of two or more metals.)
5. B) toxicity
6. No. Matter is neither created nor destroyed during a chemical reaction; it is only rearranged. The log, when burned, forms carbon dioxide and water vapor. The law of conservation of mass is upheld.
7. A) its kinetic energy increases.
8. B) ice cube melting (Living cells change through biochemical reactions. So, physical activity that changes muscle and fat is causing a chemical change. “Physical” has different meanings to a gym teacher and a chemist.)
9. True
10. E) a pure compound.
11. A) an exothermic chemical change.
12. D) a boulder resting on top of the hill

CHAPTER 3

Fundamental Measurements

ve you ever said "It took me an extra forty-five minutes to get to school this morning thanks to that **#*! construction ject"? Your statement contains two types of information: *quantity*, 45, and *unit*, minute. Would you have said "It took an extra forty-five to get to school"? Forty-five what? Dollars?

asurement systems are devised to communicate the quantity being measured and the units of that measurement. The *label* noting the unit informs us of what property is being measured, and should also convey an idea of the size of the unit. The it in your complaint is the minute. The minute is a measure of time. You expect others to know the size of the minute.

e names of units, their sizes, and the relationship s between them are described in a standard defining each measuring tem. Units have been made up and defined by people, for the purpose of measuring. For example, Congress has declared t 100 cents will comprise one dollar of American currency. That is the only reason that this relationship exists.

e measurements of science are generally of the same type as those of everyday life: mass, energy, volume, etc. Science g ago adopted the metric system, officially the SI system, which you will study in this chapter. Most of the developed rld has adopted the metric system. About 30 years ago, a metrication committee was set up to switch the U.S. to the tric system. Cooking recipes were written in liters and grams rather than cups and ounces. Large U.S. corporations that l internationally adopted the SI system, but the U.S. citizen did not. Remnants of the metrication attempt survive on oking measuring cups, on car speedometers and on other devices which indicate in both systems.

you go through this chapter, be VERY CAREFUL to pay attention to BOTH parts of a measurement. If any one part is ssing, measurement information becomes meaningless and confusing!

KILLS TO ENHANCE SUCCESS IN THIS CHAPTER

e skills required for this chapter involve the use and the interpretation of measurement quantities.

merical problems can be solved using the technique known as dimensional analysis, by using a formula, or by setting up a io. It is worth your time to become very familiar with dimensional analysis and to learn to use it. It is not a substitute for derstanding, but it can be a very powerful tool in your problem-solving arsenal.

velop a habit of estimating what a reasonable answer to a conversion problem should be. Should the answer be higher or ver than the starting value? Compare the size of the units being interconverted. Use your calculator, but remember that the culator is a tool; it's only as smart as the person pushing its buttons. Uninformed reliance on the calculator display can ult in your presenting answers that are complete nonsense.

HAPTER 3 PROBLEMS

early sections, numerical answers are given as calculated and also rounded to the proper number of significant figures, ng scientific notation as necessary. In later sections, you will be expected to apply your learning and report numerical wers with the proper number of significant figures automatically.

a Calculate the grams of sugar in a ten pound bag. Does your answer make sense?
Given: 10. lb
Need: g
Connecting Information:

454 g = 1 lb   Calculate: $10 \text{ lb} \times \frac{454 \text{ g}}{1 \text{ lb}} = 4540 \text{ g}$ Round: 4.5×10^3 g

Estimate: There is a little less than 500 grams in one pound. Therefore in ten pounds (ten times that amount), the is a little less than 5000 grams. So, our answer makes sense.

3.1b. A jumbo paper clip measures 5.0 centimeters in length. How long is the paper clip in millimeters? Estimate length in inches.

3.2a. A dose of inhaled medication provides 0.25 mg of effective ingredient. What is the mass of effective ingredient kilograms?

Given: 0.25 mg

Need: kg

Connecting Information:

1 g = 1000 mg

1 kg = 1000 g

$$0.25 \text{ mg} \times \frac{1\text{g}}{1000 \text{ mg}} \times \frac{1 \text{ kg}}{1000 \text{ g}} = 0.000\,000\,25 \text{ kg} \qquad \text{or, } 2.5 \times 10^{-7} \text{ kg}$$

3.2b. Your chemistry lecture takes 50. minutes. How long is your lecture in years?

3.3a. How many cubic centimeters are in 5.76 liters?

Given: 5.76 L

Need: cm^3 or cc

Connecting Information:

1 L = 1000 cm^3

$$5.76 \text{ L} \times \frac{1000 \text{ cm}^3}{1 \text{ L}} = 5760 \text{ cm}^3 \qquad 5.76 \times 10^3 \text{ cm}^3$$

3.3b. How many kilometers in 1256 meters?

3.4a. How many milligrams in a block of ice that has mass of 150 grams?

Given: 150 g

Need: mg

Connecting Information:

1 g = 1000 mg

$$150 \text{ g} \times \frac{1000 \text{ mg}}{1 \text{ g}} = 150{,}000 \text{ mg} \qquad 1.50 \times 10^5 \text{ mg}$$

3.4b. A quarter-pound hamburger has a mass of 114 grams. What is its mass in kilograms?

3.5a. What is the length of a 100 yard football field in kilometers?

Given: 100 yd

Need: km

Connecting Information:

1 m = 39.37 in

1,000 m = 1 km

12 in = 1 ft

3 ft = 1 yd

$$100 \text{ yd} \times \frac{3 \text{ ft}}{1 \text{ yd}} \times \frac{12 \text{ in}}{1 \text{ ft}} \times \frac{1 \text{ m}}{39.37 \text{ in}} \times \frac{1 \text{ km}}{1000 \text{ m}} = 0.0914 \text{ km}$$

3.5b. You might get a speeding ticket for driving at 70 mph on an Interstate highway where the limit is 55 mph. Wha your speed in kilometers per hour (kph)?

3.6a. You measure the length of one side of your house as 75 feet 3.5 inches. What is the precision of y measurement? The accuracy? The uncertainty?

Answer: You would probably use a steel tape of length 100 feet, a readily available device, commonly marked in 1/8 inch increments. The precision of the tape is expressed as measuring to the nearest eighth inch. An eighth inch is 0.125 inches so, in theory, you could measure three decimal places.

But, the actual precision of your measurement depends upon many other factors. If you lay the tape on the ground, the tape goes up and down over irregularities on the surface. If you have a buddy hold the end of the tape against one corner of the house and you stretch the tape to the other end, it may droop. The edge of the house is irregular and bumpy. You probably can only align the tape to within a quarter inch. Houses are usually not perfect; they may vary 1/8 inch or more from the vertical. There are many sources of imprecision.

Precision refers to *reproducibility* of a measured value. If you remeasured your house a week later using the same tape, would you obtain *precisely* the same value of its length? Even if you carefully used the same technique, a temperature change may have lengthened or shortened the tape. The house itself may have changed length.

Mass-produced devices are not *calibrated* against a standard during manufacture. The accuracy of the tape itself, as manufactured, is probably no greater than to the nearest eighth inch. The size of the smallest increment or notch is not necessarily a statement of accuracy, but without facts it is a reasonable assumption.

The total uncertainty of your measurements is the sum of your imprecisions and inaccuracies.

Error in tape alignment on one house corner:	1/4 inch
Error in tape alignment on other house corner:	1/4
Error due to tape droop, estimated:	1/2
Error from non-vertical sides	1/8
Error in reading tape:	1/8
Accuracy of tape itself:	+ 1/8
Total uncertainty:	1-3/8 inch

Does this uncertainty surprise you? Let's look at it as a percentage. Take the total uncertainty and divide by the measurement, then change to a percent.

Convert 1-3/8 inch to decimals:

$$1\text{-}3/8 = 1 + \frac{3}{8} = 1 + .375 = 1.375 \text{ inch}$$

Convert 75 ft 3.5 inch to decimals:

$$(75 \text{ ft} \times 12 \text{ in/ft}) + 3.5 \text{ in} = 903.5 \text{ inch}$$

Percentage uncertainty:

$$\frac{1.375 \text{ in}}{903.5 \text{ in}} \times 100\% = 0.15\%$$

b. Estimate the uncertainty of your house measurement if you used a 50 foot steel tape of the same characteristics as the 100 foot one. Calculate the percentage of uncertainty.

a. Express the length of the house of problem 3.6 in decimals to the proper number of significant figures.

Answer: The uncertainty in measurement is, in decimal feet:

$$1.375 \text{ inch} \times \frac{\text{ft}}{12 \text{in}} = 0.1146 \text{ feet}$$

So, the house length of 75.292 feet has uncertainty in the tenths-foot place. Therefore, only three significant digits are warranted in the measurement: 75.3 feet.

"Soft" conversions of dimensions between measurement systems can cause severe problems in manufacturing. The conversion may increase or decrease the interpreted precision required.

Note: The degree of uncertainty in the last significant figure depends ultimately upon the instrument used in measuring. A weighing on an analytical balance is to the nearest tenth milligram. The last significant figure in the weight represents its value ± 1 tenth milligram. It cannot be neglected.

b. For an experiment in the chemistry laboratory, you record the following weights on a student balance:

Weight of beaker plus chemical:	143.7 grams
Weight of empty beaker:	89.7 grams

How many significant figures are in the weight of the chemical? What is the precision of your weighing? What is the accuracy?

3.8a. Nails are sold by weight. The term *penny,* signified by the letter d following a digit, was originally an Eng weight measure. Until recently, nails were shipped in wooden kegs holding 100 pounds of nails. Some retailers buy nails in bulk and sell them by the pound from a bin.
Say you need a lot of small nails, 4d ones. You would like to know how many fourpenny nails are in a pound. You weigh several one-pound samples of 4d nails, counting the number nails in each sample. Your samples v from 57 to 63 nails. The average is 60 nails, with two significant digits.
How many nails are in 100 pounds of 4d nails?
Number of 4d nails in 100 pounds: = 57 to 63 nails/pound × 100 pounds = 5700 to 6300 nails
Only two significant figures are warranted. So, the averaged answer is 6000, or 6.0×10^3 with two sig figs.

3.8b. How many nails are in 10 pounds of 4d nails?

3.9a. The density of automotive antifreeze is 1.1 g/mL. If the radiator in your automobile holds 15 L, by how much the mass of the car increase when the radiator is filled with antifreeze vs. when the radiator is filled with water?
Given: 15 L
1.1 g/mL antifreeze
Need: g
Connecting Information:
1.0 g/mL water
1000 mL = 1 L
d = density, g/mL
m = mass, g or kg
V = volume, mL or L
Solution Method 1:

$$d = \frac{m}{V}$$

Solving the equation for m: $m = dV$

$$\text{Antifreeze: } m = \frac{1.1\ g}{1\ mL} \times 15\ L \times \frac{1000\ mL}{1\ L} = 16{,}500\ g$$

$$\text{Water: } m = \frac{1.0\ g}{1\ mL} \times 15\ L \times \frac{1000\ mL}{1\ L} = 15{,}000\ g$$

$$\text{Mass increase} = \text{Difference} = 16{,}500\ g - 15{,}000\ g = 1{,}500\ g = 1.5\ kg$$

Solution Method 2:

$$15\ L \times \frac{1000\ mL}{1\ L} \times \left(\frac{1.1\ g}{1\ mL} - \frac{1.0\ g}{1\ mL}\right) = 1500g = 1.5\ kg$$

If we had not converted to kilograms, we would have expressed our answer as 1.5×10^3 g. Note: You should n use plain water in your radiator!

3.9b. Using the data of this table, find the density of an unknown liquid:

mass of liquid, g	volume of liquid, mL
10.4	8.0
13.6	10.5
15.1	11.6

3.10a. What is your body temperature in Celsius?
Given: 98.6˚F Need: ˚C Connecting Information: $°C = \frac{°F - 32}{1.8}$ $°C = \frac{98.6 - 32}{1.8} = 37.0°C$
Note that the conversion values of 32 and 1.8 are exact, so three significant figures are still appropriate.

3.10b. Water freezes at 0˚C. What is this temperature in ˚F? (Be sure to give the complete solution to this problem if already know the answer.)

3.11a. If you drink a 150 mL cup of tea that is at 90˚C, how much heat in Joules does your body absorb? Consider th to be equivalent to water.

Given: Tea: 150 mL
T_i: 90°C Initial temperature
Need: Joules
Connecting Information:
$T_f = 37°C$ Final temperature, 'normal' body temperature
sp ht (H_2O): 4.184 J/g°C Specific heat of liquid water
d (H_2O): 1 g/mL Density of liquid water
$q = m(\Delta T)(\text{sp ht})$

$$q = 150 \text{ mL} \times \frac{1\text{g}}{1\text{mL}} \times (90\text{-}37)°\text{C} \times 4.184 \text{ J/g}°\text{C} = 33{,}262.8 \text{ J} = 3.3 \times 10^4 \text{ J}$$

Scientific notation is used to limit the answer to the warranted two significant figures.

1b. 100 mL of water and 100 mL of ethanol are each heated from 20°C to 50°C. Which absorbs more heat, and how much more? (Ethanol specific heat = 0.586 cal/g°C)

swers to "b" Problems

). 50. mm. Estimated: 2 inches.
). 0.000095 year
). 1.256 km
). 0.114 kg
). 113 kph

). You would measure 50 ft from one end, and mark there. Then you'd measure the remaining length from this mark. The uncertainties for the first 50-foot measurement are:

Error in tape alignment on house corner:	1/4 inch
Error due to tape droop, estimated:	3/8
Non-verticality of walls:	1/8
Error in reading tape:	1/8
Accuracy of tape:	1/8
Uncertainty, first measurement:	1 inch
Uncertainty of final measurement:	
Uncertainty of two measurements:	2 inch
Precision of alignment of tape at 50 foot mark, est.:	1/4
Total uncertainty:	2-1/4 inch

Percentage uncertainty $= 100 \times \frac{2.25}{903.5} = 0.25\%$ (as compared to 0.15% using a 100-ft tape)

). Weight of chemical: 143.7 grams − 89.7 grams = 54.0 grams.
Significant figures in weight: 3
Precision of weighing: To the nearest two tenths of a gram.
Accuracy of weighing: To the nearest two tenths of a gram.
The precision is to the nearest two tenths of a gram because two values were measured then combined. Equating the accuracy to the precision assumes the balance to be in good condition, zeroed, and calibrated to weigh to the nearest tenth gram, as indicated on the balance beam.

). Number of 4d nails in 10 pounds: 6.0×10^2
As written, the number of nails in 10 pounds seems merely 10 times the number of nails in 1 pound. But think of the scale used for weighing as precise to so many *average* 4d nails, or having a sensitivity of this number of nails. This precision remains the same whether 1 pound or 10 pounds is being weighed.

). 1.3 g/mL
)b. °F = 1.8°C + 32. Answer: 32°F
1b. Water absorbs 1.2×10^3 cal more than the alcohol absorbs.

PRACTICE TEST I

1. Which number in each pair should be considered an exact number?
   a. 12 in/ft 9 ft/sec
   b. 1000 mL/L 0.825 g/mL
2. To one significant figure, what would be the approximate density of each of the following?
   a. air
   b. water
   c. a single crystal of salt
3. How many significant figures should be expressed in the product of 27.633, 0.5, and 1257.7855 ?
4. What is the technical source of the limitation on significant figures?
5. What really happens when you use a thermometer to measure temperature? How does it work?
6. What three quantities determine the total amount of heat that a hot liquid can transmit to you if you place your hand in
7. Imagine that you are holding a drinking glass in one hand and a Styrofoam cup in the other. I offer to pour the s amount of boiling water in both cups. What will your response be, and why would you react differently to the two cups?
8. What is the numerical value of $\frac{\text{cubic cm}}{\text{cm}^3}$?
9. What is the product of 1×10^6 and 2.0×10^{-5}?
10. What is wrong with expressing the density of a material as $\frac{6.5 \text{ g}}{2 \text{ mL}}$?
11. If a "perfectly accurate" measuring tape could be made, would there still be uncertainty in a measurement taken with

PRACTICE TEST II

1. Which of these has the most significant figures?
   A) 0.000450
   B) 0.000435
   C) 12000
   D) 0.5008
   E) 0.23600
2. Which of these is a measurement of density?
   A) 45.6 cm B) 10.6 mL C) 13.6 g/mL D) 4.00 $/yr E) 44 ft/sec
3. Which of these measurements is the same as 100 mL?
   A) 0.1 qt B) 1 L C) 100 cm^3 D) 100 g E) 0.01 L
4. How many milligrams are in fifty kilograms?
5. Which of these containers will be about filled with 250 mL of liquid?
   A) a coffee cup B) a gallon milk bottle C) a five gallon bucket D) a 5 cc syringe
6. A calorie is 4.184 J. What is the caloric content in Joules of a 110 Calorie serving of breakfast cereal?
7. One thousand grams of water at 100 °C can transfer how much heat to your body if your body temperature is 37 °C?
8. Why does ice float on water?
9. A sample of a solid has a density of 1.50 g/mL and has a volume of 15.0 mL. What is the mass of the sample?

If the density of a sample of steel is 8.7 g/cm^3, what is the volume of 454 g of the metal?

How many significant figures in the expression 7.0050?

The correct answer to 25.53 + 16.554 + .1105 + 6.01 is _________.

The product of 6.0×10^5 and 3.11×10^{-15} is _________.

If the atmospheric temperature today is 87°F, what would it read in Celsius?

A standard piece of paper is 8.50 inches by 11.0 inches. What are the dimensions in metric units?

How many calories of heat are transferred to your hand if 500 grams of water at 95°C spills on your hand that is at 35°C?

If 3500 Cal is equivalent to one pound of body fat, how much weight can you gain from a cookie that has 150 Calories?

How much will the temperature of 15 grams of aluminum be raised by the addition of 1000 Joules of heat?

Standard room temperature is 25°C. What is this temperature in Kelvin?

At a certain temperature the density of water is 0.9867 g/mL and the density of mercury is 13.578 g/mL. What is the cific gravity of the mercury?

Answers to Practice Test I

1. The first number in both cases is the exact number, because they are conversions. The second number is a measuren of velocity in a and density in b. Conversion factors are exact. Measurements have inherent error.

2. 10^{-3} g/cm^3 (1 g/L); 1 g/cm^3; 10 g/cm^3

3. Only one, due to the limiting effect of 0.5 which has 1 sig fig.

4. The precision of measuring devices.

5. Thermometers use an alcohol or mercury reservoir, which is placed where the temperature is to be taken. Gain or los heat causes an expansion or contraction of the reservoir liquid, the final level of which is read on a scale.

6. The amount of liquid (mass), the temperature difference between the liquid and your hand, and the amount of hea liquid contains.

7. If you held the cups for a long time, the same amount of heat would eventually be transmitted to your hands from e cup (neglecting losses to the air). But Styrofoam is a better insulator than glass, so its heat would be transferred much n slowly. It would be much more comfortable.

8. One. The numerator and denominator are two ways of writing the same unit.

9. 20 or 2×10^1

10. Density, by convention, is referenced to one unit of volume. It is necessary to divide the numbers to "reduce" the rat

11. Yes, there will be uncertainty. Review Problem 3.6. Human error is always a factor.

Answers to Practice Test II

1. E
2. C
3. C
4. 5×10^7 mg
5. A
6. 4.6×10^5 cal (Remember, one "diet" big C Calorie equals 1000 "little c" calories.)
7. 63,000 cal
8. Ice is less dense than water.
9. 22.5 g
10. 52 cm^3
11. five significant figures
12. 48.20 (Round once only!)
13. 1.9×10^{-9}
14. 31°C
15. 21.6 cm by 27.9 cm
16. 3.0×10^4 cal
17. 0.043 pound
18. 74°C
19. 298 K
20. 13.76

CHAPTER 4

Elements, Atoms, and the Periodic Table

According to the Food and Drug Administration, a food is "pure" if it contains only substances that are accepted to be beneficial. You learned in chapter 2 that both elements and compounds are *pure* substances. In chemistry, a pure material is one with very definite properties, both physical and chemical. It is not really necessary to include the word pure as an adjective modifying the word substance: All *substances* are pure. Again we have a common term which, when used in science, is more specific.

Ultimately, we define a substance as a material of definite and constant composition. This distinguishes a substance from a mixture, the composition of which may vary. No matter where or when a substance is found in nature, or no matter where or when it's synthetically produced, it is exactly the same substance and has the same properties. Water collected on Mars could not be distinguished from water on Earth.

Early chemists could prove that many substances were compounds, by heating until the compounds decomposed into "simpler" substances. These simpler substances, and other chemicals that could not be decomposed, might be elements – or they might not. However, by correlating thousands of experiments, they became more confident that certain elements had been identified because they kept showing up as decomposition products. They became nearly sure of themselves when Dalton proposed his theory. He explained the results of these experiments by describing the joining of atoms to form molecules.

A list of elements and a system showing relationships between them were sorely needed. Such a system with strong relationships could help categorize elements. Then one could look at a substance and see whether it fit the category or not. If the substance did not fit, it was probably a compound, not an element. Many systems were tried in the early 1800s. Finally, in 1869, Russian chemist Dimitri Mendeleev produced a **periodic table**. This was the system that chemists needed. We still use Mendeleev's table today, modified, but its concept is essentially the same.

SKILLS TO ENHANCE SUCCESS IN THIS CHAPTER

To visualize a world beyond our senses is a skill extremely important in this chapter and from now on. Of the many worlds beyond our senses, we are interested in the world of the atom. It is a world of particles far smaller than those that we can directly see, even when using an optical microscope.

It is reasonable to assume that atoms exist. We conceptually continue to divide an object until we eventually reach a tiny particle that is the smallest particle of the object - a brick or building block of the object. We can conceptually rebuild the object from its bricks. However, if we divide this smallest particle, we no longer have a brick of the object. We have something else. We can no longer rebuild the object. The ancient Greeks thought this way. The term "atom" is taken directly from the Greek.

We cannot see these atoms. We cannot handle them. Our unaided senses give us no clue as to their existence. You may have difficulty visualizing these atoms at first Keep at it. Try using a model set to help, or even small objects of different colors.

CHAPTER 4 PROBLEMS

4.1a. Distinguish between an element and a compound.
Answer: An element is the "simplest" form of a substance, made up of only one kind of atom. It cannot be chemically broken down into other substances. A compound is formed from two or more elements. As such, it is not as simple as an element, and it can be decomposed into its component elements.

4.1b. Distinguish between an element and an atom.

4.2a. Distinguish between an atom and a molecule.
Answer: An atom is a single individual building block, the smallest unit of matter that cannot be further divided by chemical means. A molecule is formed from two or more atoms, chemically combined.

4.2b. Distinguish between the terms symbol and formula, as used by chemists.

4.3a. Oxygen in its elementary form is abundant in our atmosphere. Why was the element oxygen not discovered until 1774?
Answer: One cannot readily distinguish oxygen in the air. It is colorless, as are the nitrogen and other gases in air. Early investigators called any colorless gas "air". The alchemists discovered mercury when they heated an ore which we now know to be a compound of mercury and oxygen. Oxygen was given off during heating of the ore, but the alchemists let it escape as an air. Even Priestley, who is given credit for discovering oxygen, first called the gas "perfect air".

4.3b. Why is gold used for coinage?

4.4a. In what area of the periodic table are the structural metals located?
Answer: Our primary structural metal is iron, used in the form of steel. Aluminum is increasing in importance with time. It is used in aircraft and is being used more and more in automobiles to reduce weight. These elements are located in the metals, in the center of the table to the left of the zigzag line.

4.4b. In what area of the periodic table are the "elements of life" located?

4.5a. You know that the main processor and the memory units, or chips, in your computer are built of silicon. For what reason is this element used?
Answer: Silicon is a metalloid. The metalloids are situated between the metals and the nonmetals on the periodic table. They are near the zigzag line separating the metals from the nonmetals. The basic building block of a computer chip is a *semi*conductor. The term itself implies a substance with properties between a conductor (metal) and an insulator (nonmetal).

4.5b. What other element could be used as a semiconductor? Hint: Use the periodic table.

4.6a Two samples of water from different sources were both found to have the following mass ratio of hydrogen to oxygen:
2 g hydrogen : 16 g oxygen. What law does this exemplify?
Answer: The law of definite proportions is indicated. This is often called the law of constant composition. The two different samples of water both have the same composition.

4.6b. Another compound of hydrogen and oxygen is analyzed to have this ratio of hydrogen to oxygen:
2 g hydrogen : 32 g oxygen.
Compare this to the result of 4.6a. What law does this indicate? Show how you arrive at your answer.

4.7a. A compound of nitrogen and oxygen contains 46.7% nitrogen by weight. Another compound of these two gases contains twice as much oxygen. What is the percent by weight of nitrogen in the second compound?
Answer: Apply the third and fourth principles of Dalton's theory.

Percent oxygen in first compound: $100\% - 46.7\% = 53.3\%$

Ratio of oxygen to nitrogen in first compound: $\dfrac{53.3\%\ \text{oxygen}}{46.7\%\ \text{nitrogen}} = 1.14$ or 1.14 parts O : 1 part N

Ratio of oxygen to nitrogen in second compound (with twice as much oxygen): $1.14 \times 2 = 2.28 : 1$

Percentage nitrogen in second compound: $100\% \times \dfrac{1 \text{ part nitrogen}}{(2.28 + 1) \text{ parts total}} = 30.5\%$

4.7b. A compound contains 27.3% carbon by weight, the rest oxygen. A second compound of these two elements contains half as much oxygen. What is the percent by weight of carbon in the second compound?

4.8a. An element has a mass number of 11. It has 5 protons in the nucleus of each of its atoms. Where is its location on the periodic table? What element is it?
Answer: The presence of 5 protons in the nuclei of the element's atoms means that the element has an atomic number of 5. The element fits at the top of the periodic table. The element is boron, B.

4.8b. An element has a mass number of 73. It has 32 electrons surrounding the nucleus of each of its atoms. Where is its location on the periodic table? What element is it?

4.9a. How many are there each of electrons, protons, and neutrons in each atom of this isotope: ${}^{235}_{92}X$? What element does the notation represent?
Answer: An atomic nucleus of the element contains 92 protons and 235 − 92 = 143 neutrons. The atom (assumed to be neutral) also contains 92 electrons. The element is element number 92, uranium, U.

4.9b. An isotope of element number 38 has a mass number of 84. How many of each subatomic particle type are in each of its atoms? Identify the element.

4.10a. The atomic mass of oxygen is given as 16.00 in the periodic table. Does this mean that naturally occurring oxygen contains but one isotope, of mass number 16?
Answer: Without specific information concerning isotopes of oxygen, one would conclude that probably isotopes of mass number other than 16 exist in very low proportions. It is also distinctly possible that 2 isotopes, of mass number 15 and 17, exist in 50:50 proportions. Actually, oxygen has known isotopes of mass number 16, 17, and 18. The isotopes of mass number 17 and 18 exist in extremely low proportions.

4.10b. The atomic mass of bromine is given as 79.90 in the periodic table. Does this mean that naturally occurring bromine is composed mostly of an isotope with a mass number of 80?

4.11a. You are given 1.0 mole of iron and 2.0 moles of sulfur to react together in the laboratory. How many atoms of each element do you have?

Answer: Avogadro's number gives us the number of particles in a mole.

Atoms of iron $= 1.0 \times 6.022 \times 10^{23} = 6.0 \times 10^{23}$ atoms

Atoms of sulfur $= 2.0 \times 6.022 \times 10^{23} = 12.04 \times 10^{23}$ atoms or 1.2×10^{24} atoms

4.11b. You are given 32.1 g of sulfur and 78.2 g of potassium to react together in the laboratory. How many moles of each element do you have? How many atoms of each?

4.12a. Calculate the number of moles in 29.17 grams of magnesium hydroxide, $Mg(OH)_2$, the compound in milk of magnesia.

Given: 29.17 grams magnesium hydroxide

Need: moles

Connecting Information:

1 mole = formula weight in grams

Calculate the formula weight of magnesium hydroxide.

$$\text{F.W.} = 24.31 + 2(16.00 + 1.008) = 58.33 \frac{\text{g}}{\text{mol}}$$

Calculate the number of moles of magnesium hydroxide.

$$\# \text{ moles} = 29.17 \text{ g} \times \frac{1 \text{ mole}}{58.33 \text{ g}} = 0.5000 \text{ mole}$$

4.12b. Calculate the number of moles in 143.1 g of sodium carbonate decahydrate, $Na_2CO_3 \cdot 10H_20$, washing soda.

Answers to "b" Problems

4.1b. An element is a substance in the macroscopic world in which we live. We can handle a mass of an element. An atom is a building block of an element. It is the smallest particle within a sample of element that still has the essential properties of the element. The atom exists, but in an invisible microscopic world.

4.2b. A symbol represents an element or an atom; a formula represents a compound or a molecule.

4.3b. Gold occurs in nature as an element. It is readily recognizable, is soft and shiny, and can be easily worked into intricate shapes that do not tarnish or rust. Gold coins can be easily stamped, shaped, or engraved. You know that jewelry, crowns and many artifacts have been made from gold for thousands of years.

4.4b. The elements of life are primarily carbon, oxygen and nitrogen.
These elements are located in the center of the nonmetals, at the top of the table.

4.5b. Look at the periodic table and pick an element with properties similar to those of silicon.
Either germanium, Ge, or carbon, C. Both have properties similar to silicon.
Early semiconductors were made of Ge. C is potentially useful, but has been without experimental success.

4.6b. The law of multiple proportions. Water had a 2:16 ratio. This compound has a 2:32 ratio, or double the amount of oxygen. Whole-number multiples indicate multiple proportions. [The second compound is hydrogen peroxide.]

4.7b. Percent carbon in first compound: 27.3% Percent oxygen in first compound: 72.7%

Ratio of oxygen to carbon in first compound: $\frac{72.7\%\ \text{oxygen}}{27.3\%\ \text{carbon}} = 2.66$ parts oxygen : 1 part carbon

Ratio of oxygen to carbon in second compound (with half the oxygen): $2.66 / 2 = 1.33$ parts O : 1 part C

Percentage carbon in second compound: $100\% \times \frac{1\ \text{part carbon}}{(1.33 + 1)\ \text{parts total}} = 42.9\%$

4.8b. The element has 32 protons in the nucleus of each atom. The atoms are neutral. The number of protons equals the number of electrons in each atom. The atomic number is 32. The element is germanium, Ge.

4.9b. The atom contains 38 protons, 38 electrons and 46 neutrons. The element is strontium, Sr.

4.10b. Bromine is primarily composed of two isotopes of mass number 79 and 81, which exist in nearly a 50:50 ratio. The isotope of mass number 79 is slightly more abundant.

4.11b. 1.00 moles of sulfur (6.02×10^{23} atoms) and 2.00 moles of potassium (1.20×10^{24} atoms)

4.12b. 0.5000 mole

PRACTICE TEST I

1. Who was the "father" of quantum mechanics, the theory that led to our current model of the atom?
2. What is the <u>second</u> most common element in the universe?
3. Based on its location in the periodic table, what type of element is mercury (Hg)?
4. Give the chemical symbol and name of any nonmetal.
5. Which one of the following is ductile? C Ag He As S
6. List the formulas of all the elements that are diatomic gases at room temperature.
7. Who is credited with discovering the atom?
8. What is the name of the positively charged subatomic particle?
9. How many protons, neutrons, and electrons are in the radioactive isotope carbon-14, used to date archeological artifacts?
10. What atom has 16 neutrons and a mass number of 32?
11. Calculate the average atomic weight of neon, given the following data:

^{20}Ne	20.0 amu	90.92 %
^{21}Ne	21.0 amu	0.250 %
^{22}Ne	22.0 amu	8.83 %

12. How many moles are there in a sample of carbon containing 9.03×10^{24} atoms?

13. How many <u>atoms</u> of hydrogen are in 0.250 moles of hydrogen?

14. What is the molar mass of calcium carbonate ($CaCO_3$), a major component of limestone?

15. What is the mass of 0.50 moles of carbon dioxide, CO_2 ?

16. You buy a 25 pound bag of charcoal (carbon). How many moles of carbon are present in the bag?

17. How many carbon atoms are in your bag of charcoal?

18. How many oxygen atoms are present in 100. pounds of calcium carbonate, $CaCO_3$?

PRACTICE TEST II

1. What is the chemical symbol for lead?

2. What is the most abundant chemical element in the universe?

3 Which of the following is a metalloid? Ar Al Au S Se

4. Nitrogen is classified as what type of element?

5. Which of these elements would you expect to be shiny? S C Kr Na P

6. Name an element that is a liquid at room temperature.

7. What is the law that holds that a given compound always contains certain elements in a fixed proportion, regardless of origin?

8. Give the name of the nonnuclear subatomic particle.

9. What is the name of the element with twenty protons in the nucleus?

10. Fallout from atom bombs contains the deadly nuclide ^{90}Sr (strontium-90). How many neutrons does ^{90}Sr have?

11. A certain element loses an electron to form an ion with +1 charge. The ion formed has 60 neutrons and 46 electrons. What is it?

12. An alpha radiation particle consists of a helium-4 nucleus only, represented as ^{4}He. If an atom of radium-226, ^{226}Ra , gives off one particle of alpha radiation, ^{4}He , what is left in the nucleus afterward? (Hint: Count protons and neutrons present before and after the alpha particle is emitted.)

13. What is the average atomic weight of boron, as calculated from the isotopic abundance data given below?

^{10}B	10.02 amu	18.83 %
^{11}B	11.01 amu	81.17 %

14. What is the molar mass of chlorine?

15. What is the molar mass of ammonium sulfate, $(NH_4)_2SO_4$?

16. How many particles are there in a million moles?

17. How many moles are present in 1.2×10^{25} atoms of carbon?

18. Natural gas, which we use for cooking and Bunsen burners, is mostly methane, CH_4 . How many grams are there in 2.50 moles of methane?

19. How many moles are present in 35.5 grams of sodium sulfate, Na_2SO_4 ?

20. Since this is an exam, you may be experiencing stress, which causes the body to produce adrenaline, $C_9H_{13}NO_3$. If you produce 0.050 mg of this powerful hormone, you will be panting and your heartbeat will quicken. How many molecules of adrenaline have rushed into your bloodstream?

Answers to Practice Test I

1. Schroedinger
2. helium (He)
3. Mercury is a metal.
4. For example, C, carbon N, nitrogen O, oxygen S, sulfur, etc.
5. Ag (silver) is ductile, as a metal.
6. H_2, N_2, O_2, F_2, Cl_2 are gaseous.
7. Dalton
8. proton
9. 6 protons, 8 neutrons, and 6 electrons
10. sulfur-32 (^{32}S)
11. 20.2 amu (3 significant figures)
12. 15.0 moles
13. 3.01×10^{23} atoms (in 0.250 moles of diatomic H_2)
14. 100. g/mole
15. 22 g
16. 950 moles (945.8 rounded to 2 significant figures)
17. 5.7×10^{26} atoms (give or take a trillion or so)
18. 8.20×10^{26} oxygen atoms

Answers to Practice Test II

1. Pb
2. H
3. Se
4. nonmetal
5. Na
6. bromine or mercury
7. the Law of Definite Proportions, or the Law of Constant Composition
8. electron
9. calcium
10. 52 neutrons
11. $^{107}Ag^{+}$
12. ^{222}Rn Subtract out alpha's subatomic particles:

$$\begin{array}{lrrr} & ^{226}Rn = & 88\,p & 138\,n \\ - & ^{4}He = & -2\,p & -2\,n \\ \hline = & ^{222}Rn & 86\,p & 136\,n \end{array}$$

13. 10.824 amu

$$\begin{array}{r} 10.02 \text{ amu} \times .1883 = 1.887 \\ 11.01 \text{ amu} \times .8117 = \underline{8.937} \\ 10.824 \end{array}$$

(See rules for addition and significant figures.)

14. 70.9 g/mole (Remember, chlorine is diatomic Cl_2.)
15. 132 g/mole
16. 6.02×10^{29} particles
17. 20. moles (2.0×10^{1} moles)
18. 40.0 g
19. 0.250 moles
20. 1.6×10^{17} molecules

CHAPTER 5

Atomic Structure: Atoms and Ions

The material in this chapter is an excellent example of the rigorous *scientific method*. The scientists who discovered the structure of the atom were physicists. They were not thinking of chemical reactions. At first, they were not even thinking of the atom. They were studying spectra. They were studying electrical phenomena. They were studying light. It all came together in the structure of the atom.

Why do chemistry texts include this physics? It is overwhelming at first. But the study of atomic structure gives chemists an understanding of chemical behavior, and helps them find methods by which to better control reactions.

SKILLS TO ENHANCE SUCCESS IN THIS CHAPTER

Two skills are now becoming critical:

- •visualization of a world beyond our senses.
- •logical inference.

These skills go hand-in-hand. It is by logical inference that we learn of the worlds beyond our senses.

Scientists *experiment* to gain knowledge of the universe and its workings. They then try to explain the results of their experiments by *theory*. What they observe is not produced by magic. There has to be a logical reason, a cause-and-effect relationship. They keep looking until they find it.

In the series of experiments described in the text, scientists were observing atomic events that indirectly communicated information to them. The existence of atoms wasn't obvious at first. Atoms were thought to be the smallest particles, indivisible and structureless. Then men like Thompson, Rutherford, and Bohr came along and saw connections between disparate experiments. They saw an underlying commonality which revealed that the atom does have a structure consisting of smaller pieces. The basic concept was easy enough to visualize: tiny planets revolving around a nuclear sun. But because these "planets" jump around between various energy levels, it was hard to locate exactly where they were. Electrons are slippery characters.

Then Pauli said that it's impossible to pinpoint where one of these tiny electrons is at any time. Einstein found the link between energy and subatomic particles. Schroedinger came up with ways to predict the likelihood of different electron locations. These scientists developed the *quantum mechanical* concept, according to which an electron rapidly moves all over the place. One could know only the *probability* of it being in a given position at a given time. Certainty was lost, and visualization became even more difficult. The electron was not an obvious neat little ball.

Try to understand what data were collected in each of the various experiments noted. Try to follow the thinking of the investigators who put the results of many experiments together into a unified set of conclusions, a theory. Don't be discouraged by the quantity of data or the difficulty of visualization. Use the periodic table to see patterns.

CHAPTER 5 PROBLEMS

5.1a. Rutherford discovered three types of radiation emitted by radioactive elements. These he called alpha, beta and gamma "rays". But we speak of light rays, infrared rays, ultraviolet rays, X rays, etc. and consider that they travel in waves. Are all of Rutherford's rays like light rays?
Answer: The three rays emitted are not of the same nature. Alpha rays and beta rays have mass and are therefore streams of particles. Gamma rays have no mass and are of the nature of light rays. See Table 5.1 in your text. Gamma rays belong to the electromagnetic spectrum, along with light rays.

The term *ray* was originally applied to any type of observed radiation, since little was known about the nature of the radiation. The terms alpha, beta, and gamma were applied arbitrarily to the emissions since they were not as yet identified as belonging to any class of phenomena or material.

5.1b. Nearly sixty years after the second world war, the developed world is still striving to prevent the use of "nuclear devices", atom bombs. Not only are these devices immediately highly destructive, but their effects linger for years, producing cancer in humans exposed to the blast or the fallout and possible genetic defects in their offspring. What does this tell you about the type of radiation from a nuclear blast? Hint: Why does the dentist cover you with a lead-containing blanket when he/she X rays your teeth?

5.2a. Tanning parlors originally used electron tubes that generated UVB rays. The idea was to emulate the sun's rays. Then tanning parlors converted to UVA rays, and promoted them as tanning without the burning and as harmless. What reasons would you have to believe that UVA rays would be harmful?
Answer: The fact that these rays will tan you proves that they are harmful. They may take longer than the UVB rays, and they will not burn you as fast, but ultimately your skin will still be harmed. Tanning is a defense mechanism mounted by the body to minimize short-term harm such as sunburn itself. Dark-skinned people have this defense already genetically built in. They can still sunburn, however.

5.2b. You cannot turn a microwave oven on unless its door is tightly latched. A screen is imbedded in the glass of the door. These features prevent microwaves from "escaping" from the oven. For what reason do we want to contain the microwaves?

5.3a. Again consider the ultraviolet spectrum. You have heard that scientists are concerned that the ozone layer high in our atmosphere is diminishing. For what reason are scientists concerned?
Answer: UVC rays are absorbed by the ozone. These rays are very dangerous to higher life. They are the highest frequency rays in the UV spectrum and therefore have the highest energy. The loss of protective ozone has resulted in a dramatic increase in skin cancer.

5.3b. Early photographic film responded much more slowly to artificial illumination than to sunlight. The film could be developed in a "darkroom" that had sufficient illumination for a person to see. The illumination used was red. Explain the slow response of the film. Hint: Consider the relationship of frequency and energy in the electromagnetic spectrum.

5.4a. How many electrons are in each of the atoms represented by these symbols?
a. Na^+ b. F^- c. Ar d. P^{3-}

Answer: Use the periodic table to find the number of electrons in each energy level:

	First	Second	Third	Total	
Na^+	2	8	0	10	Ion: One valence electron lost from third level.
F^-	2	8	–	10	Ion: One electron gained in second level.
Ar	2	8	8	18	Neutral atom.
P^{3-}	2	8	8	18	Ion: Three electrons gained in third level.

5.4b. The number of electrons in the Na^+ ion is equal to the number of electrons in what neutral atom?

5.5a. In what fundamental way does the original Bohr model of the atom differ from the quantum mechanical one?
Answer: Bohr proposed a planetary system of electrons in which the electrons were definite particles in fixed orbits around a nucleus.
The quantum mechanical concept treats the electron as existing within the atom at an energy level but without a definite orbit. Ultimately, the electron is treated as a diffuse entity whose density varies from place to place within the atom. No orbit is defined. In the original Bohr model, if one knew the *ground state* orbit of an electron and an initial position, one could calculate the position of the electron at any time in the future, as long as it stayed in the ground state. This is equivalent to our calculation of planetary positions within our solar system. In the quantum model, one can calculate only the *probability* that an electron can be found in a given position at a time.

5.5b. What property of atoms indicates the presence of discrete energy levels?

5.6a. What would be the maximum number of electrons in the seventh energy level?
Answer: $2 \times 7^2 = 98$

5.6b. A given atom has electrons in the first three energy levels. The total number of electrons in the third energy level is 3. The atom is neutral. This is an atom of what element?

5.7a. What is the Lewis dot diagram for the alkali metals?
Answer: **•X**

5.7b. What is the Lewis dot structure for F^-?

5.8a. A neutral atom has 6 electrons in its $3p$ orbitals, which are the outermost orbitals with any electrons. How many valence electrons are there in this atom? What is the element?
Answer: The atom has 8 valence electrons: 2 *s* electrons, and 6 *p* electrons in the third energy level, the outer level. Since the valence level is filled, these electrons are not available for bond formation in normal chemical reactions. The element is argon, Ar, a noble or inert gas.

5.8b. A neutral atom has five *p* electrons in its *p* orbitals. It has a total number of 17 electrons. How many valence electrons are there in this atom? What is the element?

5.9a. What empty orbitals are present in the valence level of a phosphorus atom?
Answer: The $3d$ orbitals are empty. The third principal energy level (the valence or outermost shell) is partially full. This level contains *s*, *p*, and *d* orbitals but does not contain *f* orbitals.

5.9b. In which principle energy level are the *d* electrons of silver (Ag)?

5.10a. What is the electron configuration for Ca^{2+}?
$1s^22s^22p^63s^23p^6$

5.10b. What is the electron configuration for Se^{2-}?

Answers to "b" Problems

5.1b. The blast produces high-energy "rays" of the X ray and gamma ray type. See Figure 5.15 in your text. These rays readily penetrate flesh.

5.2b. Microwaves cook food by heating it internally. They would cook your flesh in the same way.

5.3b. Of the colors in the visible spectrum, the film was most sensitive to blue light and least affected by red light. Blue light is of the highest frequency and is therefore of highest energy. Incandescent illumination is yellowish. Red light is of the lowest frequency and lowest energy, so it was used in the darkroom. The film was also sensitive to any electromagnetic radiation of frequency higher than blue light, such as ultraviolet. Pictures taken at high altitudes and on the ocean or beach would often be overexposed by UV light, which the photographer would not see and would not adjust the exposure for.

5.4b. Neon, Ne. Look at Figure 5.23 of your text.

5.5b. The occurrence of bright line spectra rather than continuous spectra indicates the quantized levels.

5.6b. Aluminum, Al

5.7b. $:\ddot{\underset{..}{F}}:^-$

5.8b. There are 2 electrons in the first level and 8 in the second, so the rest, 7, are in the third. In the third level, 2 electrons are in the *s* orbital and 5 in *p* orbitals. The atom has 7 valence electrons. The element is chlorine, Cl.

5.9b. They are in the 4th principal energy level. Silver is one of the transition elements.

5.10b. $1s^22s^22p^63s^23p^64s^24p^6$

PRACTICE TEST I

1. Which part of the e/m ratio was discovered by Millikan?
2. What property of atoms indicates the presence of a quantized system?
3. What elements could be used to produce an American flag in a fireworks display?
4. What would be the maximum number of electrons in the fourth energy level?
5. What is a major shortcoming of the Bohr model?
6. How many electrons (total) are in each of these?

   a. K^+ b. Cl^- c. Ar d. P^{3-}
7. What is the complete electron configuration for Mg^{2+}?
8. What is the orbital diagram for p^3 ?
9. Explain the order of orbital filling of a *d* sublevel.
10. What is the electron configuration in the *f* sublevel when the first electron pair appears?
11. What is the valence level electron configuration of the halogen (fluorine) family?
12. What group has the valence level configuration s^2p^4 ?
13. What would be the valence level electron configuration of element #120?
14. What is the shorthand electron configuration for Mn?
15. What is the Lewis electron-dot symbol for Cl^-?
16. What is the generic Lewis electron-dot symbol for the halogens?
17. What is the Lewis electron-dot symbol for S?
18. What type of electrons are located in a spherical orbital?
19. In which principle energy level are the *d* electrons of iron (Fe)?
20. What empty orbitals are present in the valence shell of gallium (Ga)?

PRACTICE TEST II

1. What experiment established that the atom is mostly empty space and all of its mass is in the nucleus?
2. What is the relationship between wavelength and energy?
3. What causes an atom to emit a photon of light?
4. Why does each element have its own unique emission spectrum?
5. Which type of sublevel has the lowest energy?
6. What is the electron configuration of sulfur (S)?
7. What is the orbital diagram for beryllium (Be)?
8. Draw the Lewis electron dot-diagram for francium (Fr).
9. What is the complete electron configuration for Al^{3+}?
10. Draw the complete orbital diagram for S^{2-}
11. Draw the Lewis electron-dot diagram for O^{2-}
12. How many unique electron-dot diagrams are there?
13. What is the maximum number of electrons in the third energy level?
14. How many unpaired electrons are there in As?
15. What is the maximum number of electrons in a *p* orbital?

16. What is the complete electron configuration for lead (Pb)?
17. What is the shorthand configuration for rubidium (Rb)?
18. What is the shorthand configuration for selenium (Se)?

Answers to Practice Test I

1. The charge on the electron was discovered by Millikan.
2. An atom's emission bright line spectrum demonstrates quantization.
3. We could use lithium or strontium for red; magnesium for white; copper for blue.
4. $2(4^2) = 32$
5. The Bohr model does not provide for complex electron configurations; it lacks sublevels and orbitals.
6. a. 18 b. 18 c. 18 d. 18
7. $1s^22s^22p^6$
8.  ↑ ↑ ↑
9. Place one electron in each of the 5 orbitals, with parallel (same) spins, until the sublevel is half filled. Then the electrons pair, with five more of opposite spin placed in the orbitals, until there are 10 total.
10. f^8
11. s^2p^5
12. oxygen group (VIA)
13. $8s^2$
14. $[Ar]4s^23d^5$
15. 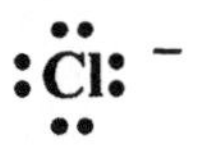 $:\ddot{\underset{..}{Cl}}:^-$
16.  $:\ddot{\underset{\cdot}{X}}:$
17.  $:\ddot{\underset{\cdot}{S}}\cdot$
18. s
19. 3rd
20. The $4d$ and $4f$ sublevels are empty. (The $4s$ and $4p$ sublevels have electrons.)

Answers to Practice Test II

1. The Rutherford gold foil experiment showed empty space in the atom.
2. They are inversely proportional.
3. An electron returning to ground state causes emission of a photon.
4. Each element has a unique electron configuration.
5. The s sublevels are lowest in energy.
6. $[Ne]3s^23p^4$
7. The orbital diagram for Be is

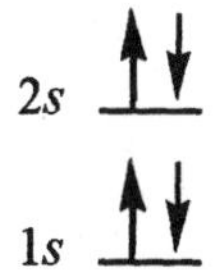

   $2s$ ↑↓

   $1s$ ↑↓
8. •**Fr**
9. $1s^22s^22p^6$
10. The orbital diagram for S^{2-} is

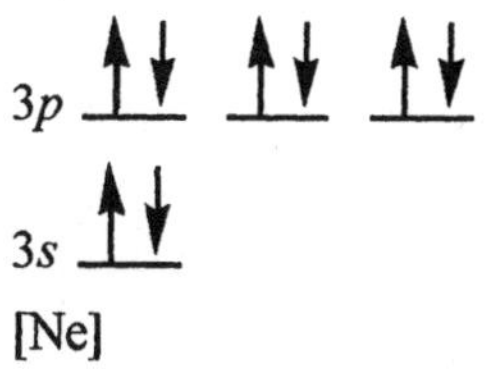

    $3p$ ↑↓ ↑↓ ↑↓

    $3s$ ↑↓

    [Ne]

11. $\left[:\ddot{\underset{..}{O}}:\right]^{2-}$

12. 8
13. $2(3^2) = 18$
14. 3
15. 6
16. $1s^22s^22p^63s^23p^64s^23d^{10}4p^65s^24d^{10}5p^66s^24f^{14}5d^{10}6p^2$
17. $[Kr]5s^1$
18. $[Ne]3s^23p^2$

CHAPTER 6

Names, Formulas and Uses of Inorganic Compounds

Infants learn to speak by first naming every object in sight. Names give us a mental handle to relate to something in the real world. We name everything so that we can communicate with others. Time was when people had only one name. For example, when we read about Leonardo daVinci, we think of daVinci as his last name. We might say the Mona Lisa was painted by daVinci. But his name was only Leonardo; "daVinci" means he came from Vinci. Many last names actually started out as birthplaces or descriptions or occupations. Nowadays we use two or three names. The more people there are, the more compound the name.

To obtain maximum order and minimize confusion, we had to systematize our nomenclature. This is extremely important when we have millions of chemicals to name. It helps us to organize our thinking and communicate more efficiently. The IUPAC (International Union of Pure and Applied Chemistry) devised a system of rules for naming chemical compounds. Engineers, chemists, and other scientists use these rules world-wide "by convention", which in this case means by mutual agreement. After you learn the system, you will be able to write the formula for a compound from its name. And you will be able to say the compound's name if given its formula.

Would you rather use names like water, lye, soda, vitriol, lunar caustic, muriatic acid, laughing gas, grain alcohol, wintergreen, and ammonia? Many of these *common* names, as they are now called, are still used, especially in commerce. If you said to the waiter or waitress in a restaurant, "I would like a glass of dihydrogen monoxide", I wonder what you would get.

SKILLS TO ENHANCE SUCCESS IN THIS CHAPTER

Learning the naming system for chemical compounds is mostly memorization. You might have to learn a little Latin and Greek, but hang in there. It won't be Greek to you after a bit.

A review of simple operations involving signed numbers might be helpful.

CHAPTER 6 PROBLEMS

6.1a. Iron is commonly listed as a nutrient on food labels. Do we ingest iron in our food? Some prepared cereals actually contain powdered metallic iron. Are you surprised?
Answer: You can digest solid metallic iron, which is Fe: uncharged iron atoms in the ground state. The hydrochloric acid in your stomach will dissolve the iron. We can "eat nails" – a very small amount – as iron powder, not as nails. Unprocessed foods do not contain metallic iron; if it is there, it has been added as a supplement. Iron found in plants and animals is in ionic form, Fe^{2+} or Fe^{3+}, as some type of dissolved salt. The metallic iron is actually less likely to cause intestinal upset than some iron additives such as ferrous sulfate. Too much iron is linked to heart attacks, especially in men. Too little causes anemia, commonly in women.

6.1b. Food labels also give the amount of sodium in the foods. In what form is the sodium, and what would be the technically correct name for the sodium in our food? A Michigan State University research group has synthesized the Na^- ion. What would be a good name for it?

6.2a. What is the formula of bicarbonate? Is there a "2" involved? How would you know, without being told or without reading it, that a bicarbonate ion is a stable chemical species?

Answer: HCO_3^-. The "bi-" prefix refers to hydrogen, not the number 2. The fact that bicarbonate has been assigned a permanent name is a clue that it is a stable, common ion. If a polyatomic ion were not stable, it would dissociate – come apart – in solution in a polar solvent like water.

6.2b. Make an educated guess: What is the formula for tungstate?

6.3a. Write the formula for calcium nitride.
Answer: Calcium nitride is an ionic compound, not an element. Ionic compounds are formed when metals bind with nonmetals, generally. The two parts of the name correspond to two different ions. Calcium is a Group II metal, so it has a +2 charge. Metals do tend to lose electrons. Nitride is a monatomic anion. Since N is in Group V, it must gain three electrons to fill its orbitals (like those of noble gases) with 8. Therefore, its charge is –3. Nonmetals tend to gain electrons.

We wish to form a neutral compound from these ions. So, to balance the positive and negative charges, we will need a least common multiple of six. We need two nitrides (–3 each) for every three calcium (+2) ions. Two nitrides means a subscript of two for nitride, and a subscript of three for calcium.
3 Ca ions @ +2 = +6 2 N ions @ –3 = –6 +6 and –6 cancel for a neutral compound.

Calcium, as cation, goes first: Ca_3N_2
No parentheses are needed, as there are no polyatomic ions. No charges show in a neutral compound. The ion charges are still there, but they cancel out to zero. The total (net) charge on the entire formula is zero.

6.3b. Titanium is a transition metal that has a +4 charge in most of its compounds. Titanium oxide is a white pigment used in paints and cosmetics. What is the formula for titanium oxide?

6.4a. Write the formula for ferrous nitrite.
Answer: Ferrous nitrite is an ionic compound, not an element. The ferrous ion has a +2 charge. Nitrite is a polyatomic anion. You should have memorized the formula for nitrate as NO_3^-. Then it is easy to remove one oxygen to get the corresponding -ite anion: nitrite is NO_2^-, with everything else (beside oxygen) staying the same. (So, memorize only the "-ates". The "-ites" are the same except with one less oxygen.) To balance positive and negative charges, we will need two nitrites (–1 each) for every one ferrous (+2) ion. Two nitrites means a subscript of two for nitrite. Any time you need more than one polyatomic ion, you must use parentheses with your subscript. Iron (the cation) goes first, then the nitrites:

$$Fe(NO_2)_2$$

Make sure that the ionic charges don't show in the final compound; they have been cancelled out. The fact that there are two twos is just a coincidence.

6.4b. Write the formula for ammonium sulfate.

6.5a. Gold chloride is commonly used in industry. It is used in solution to plate metallic gold on jewelry and on electrical contacts. Does the name definitively indicate the formula of the compound?
Answer: No, the name is indefinite, since the gold ion can have two charges, +1 and +2.
The actual name of the compound would be:
Aurous chloride or auric chloride, using the older naming convention, or
Gold(I) chloride or gold(II) chloride, using the present naming convention.
The industrial compound is actually gold(II) chloride. Most industrial compounds are those having the higher cationic charge. These tend to be more stable when exposed during handling.

6.5b. Write the name of $CuHCO_3$ in both the older system and the Stock System.

6.6a. The compound CS_2 is carbon disulfide. Should the formula actually be written S_2C and the compound called disulfur monocarbide? Is this an ionic or a covalent compound?
Answer: Carbon and sulfur are nonmetals. We would expect the bond between them to be covalent. Note the sequence of naming, as given in your text:
B Si C P N H S I Br Cl O F
The sequence tells us that the carbon should be named first, so the compound is carbon disulfide. This sequence is difficult to memorize. Let's try to figure things out. Note the position on the periodic table of

each element in the sequence relative to the zigzag line of the "staircase". In the name of a compound of two nonmetallic elements, the element nearest the staircase is first named. Carbon is closer to the stairs than sulfur. Carbon is named first.

6.6b. A compound of boron and hydrogen exists in which boron carries a formal charge of +3. Give the name of the compound. Hints: Hydrogen fits on both sides of the periodic table, and can exist as both cation or anion, depending on its partner. Formal charges can be used to find the formulas of nonmetallic binary compounds.

6.7a. Prefixes such as per– directly reference the oxidation state of a central atom. What is the oxidation state of the two elements in hydrogen peroxide, H_2O_2 ?

Connecting Information:

The molecule is neutral: oxidation states must add to zero. Total positives must equal total negatives. Take the oxidation state of H as +1.

$$2 \times (\text{oxidation state of H}) + 2 \times (\text{oxidation state of O}) = 0$$

$$2 \times (+1) + 2 \times (\text{oxidation state of O}) = 0$$

$$(+2) + \underline{2 \times (\text{oxidation state of O})} = 0$$

Therefore $\underline{2 \times (\text{oxidation state of O})}$ must equal –2 total, to cancel the +2.

The two oxygens contribute a total of –2 charge.

$$\underline{2 \times (\text{Oxidation state of O})} = -2$$

$$\text{Oxidation state of one O} = \frac{-2}{2} = -1$$

The individual oxidation state of oxygen is almost always –2 as normal oxide, O^{2-}, but in this case, oxygen takes a –1 charge. You may wonder why the formula isn't 1 : 1, written HO. The peroxide ion is polyatomic and occurs only in pairs of O^{-1} pieces: the peroxide is O_2^{2-}.

6.7b. What is the formula for magnesium dichromate, and what is the individual oxidation state of chromium in that compound?.

6.8a. Is it improper to call the pure compound H_2SO_4 by itself "sulfuric acid"?

Answer: Technically, yes, but we do it often. Acids as we define them produce the hydrogen ion, H^+, in water solution. H^+ dissolved in water is also known as hydronium ion, H_3O^+. Dissolved H^+ makes acid solutions taste sour. The hydronium (dissolved H^+) ion forms only when the compound reacts with water:

$$H_2SO_4 + H_2O \longrightarrow HSO_4^- + H_3O^+$$

For convenience, and because of old habit, we do call the compounds themselves acids. However, to be technically correct, the pure compund is hydrogen sulfate, and it is not sulfuric acid until dissolved in water and the H+ dissociates. We can also say that sulfuric acid *ionizes* in water, and write a shortcut equation:

$$H_2SO_4 \longrightarrow H^+ + HSO_4^-$$

The water is assumed to be there. "Concentrated sulfuric acid" or hydrogen sulfate is shipped in steel tank cars. Far from reacting with the steel to destroy the car, the compound coats the steel and protects it. The chemical is called "oleum" when highly concentrated. Oleum consists of excess SO_3 dissolved in liquid H_2SO_4. No water is present, therefore no dissolved H^+ ion is present, and technically it is not an acid.

6.8b. What should hydrogen perchlorate, $HClO_4$, be called in aqueous solution?

6.9a. Is the water of hydration a part of a salt molecule?

Answer: The water of hydration is bound along with the ions of a salt in a crystalline lattice. It is not a part of the basic unit of the salt. Water of hydration is present only in solid crystals. It is more like solid ice than liquid "wet" water.

6.9b. How do you name a compound that contains water(s) of hydration?

6.10a. Table 6.10 of your text indicates that "anti-acids" or antacids may be made from limestone. Are these antacids used for human consumption?
Answer: Antacids taken internally for stomach upsets contain calcium and magnesium carbonates or magnesium or aluminum hydroxides precipitated from water solution. Produced in this way, they are initially almost pure substances and composed of extremely fine particles which react readily with stomach acid. It wouldn't work very well to swallow a "rock" of limestone, which is calcium carbonate.
Calcium supplements have been made from many natural materials, including limestone, bones, and oyster shells, readily available raw materials. These are not highly refined and thus are mixtures. For example, "dolomitic" limestone also contains magnesium. Bone from cattle may contain "heavy" metals such as lead. Canned salmon contains salmon bone, which can be eaten for its calcium content.

6.10b. Why is carbon dioxide in solid form called "dry ice"?

Answers to "b" Problems

6.1b. The sodium is in the form of simple salts, such as sodium chloride and sodium bicarbonate. Most will be in the form of sodium chloride. The sodium alone should be expressed as sodium ion. It is not in elemental (zero charged) form. We do not use prefixes or endings on the names of positive ions as we do negative ions. We just add the word ion. To put this on food labels would probably confuse most people. The rare and unusual Na^- could be called "sodium anion" or it could be named as other anions are, with an -ide ending, perhaps "sodide"? Since we often revert to the Latin or Greek root of the element's name, "natride" may be the best choice.

6.2b. Tungsten is #74, with the symbol W. It is in chromium's family of transition metals. The -ate ending is used for oxyanions, those anions containing oxygen. Chromate is CrO_4^{2-}, so it is reasonable to guess that tungstate is similar: WO_4^{2-} This turns out to be true.

6.3b. TiO_2. The +4 charge of Ti must be balanced by two oxides of –2 charge each. Ti goes first, as the positive ion. Be careful not to merely "cross" the two charges, or you may get a formula of Ti_2O_4, which is wrong. This shortcut which you may have heard of has its drawbacks.

6.4b. $(NH_4)_2SO_4$. Ammonium is a +1 ion, whose formula is NH_4^+. Parentheses are needed because of the 2 subscript. Since there is only one sulfate ion, SO_4^{2-}, no parentheses are needed. The charges should not show in the final neutral compound.

6.5b. Cuprous bicarbonate or copper(I) bicarbonate.

6.6b. Boron trihydride. Hydrogen is listed after boron in the naming sequence. Both are nonmetals. There are three hydrogens, because H will have the formal charge of –1, while boron has a +3 oxidation state, formally.

6.7b. $MgCr_2O_4$. The oxidation state of chromium is +6. (Mg +2, four O's @ –2, so the two Cr's have +12.)

6.8b. Perchloric acid.

6.9b. Write the name of the salt compound. Then write "hydrate" with a Greek prefix, which indicates the number of molecules of water associated with each molecule of the compound. See Table 6.9 of your text.

6.10b. It is called dry, because it does not melt to water or other liquid, ice because it is very cold and used for cooling. The CO_2 *sublimes* directly from a solid state to a vapor state.

PRACTICE TEST I

1. What is the name of the B^{3-} ion?
2. Give the symbol for the zinc ion.
3. Combine the two ions above into a compound and write its formula.
4. Combine the aluminum Al^{3+} and sulfide S^{2-} ions into a compound and write its formula.
5. Combine nitrate ion, NO_3^-, with zinc ion into a compound and write its formula.
6. Write the formula for magnesium chloride.
7. Write the formula for lithium sulfite.
8. Write the formula for tin (II) hydroxide.
9. Write the formula for ferrous hypochlorite.

10. Name the compound $NH_4C_2H_3O_2$.
11. Name the compound KCN .
12. Name the compound $Pb(IO_4)_2$ by the old "-ous -ic" suffix system. (Hint: Oxyanions of iodine and bromine are named the same as those of chlorine.)
13. Name the compound $(NH_4)_2Cr_2O_7$.
14. Use the Stock system to name $CoCO_3$.
15. Name the compound $KMnO_4$.
16. Write the formula for the covalent compound carbon dioxide.
17. Write the formula for iodine heptafluoride.
18. Write the formula for phosphorus triiodide.
19. Name the compound N_2O_5 .
20. Name the compound SF_6 .
21. What is the individual oxidation number of Br in $KBrO_2$?
22. What is the individual oxidation number of S in $ZnSO_3$?
23. What is the individual oxidation number of C in $C_2O_4^{2-}$?
24. Name the acid HBr .
25. Name the acid H_2SO_4 .
26. What is the formula for chlorous acid?
27. Name the compound $CoCl_2{\cdot}6H_2O$, used in "weather stones".
28. What is the formula of the anhydrous salt of sodium thiosulfate pentahydrate?
29. Lye, or drain opener, consists of what chemical?
30. A whitish liquid suspension of $Mg(OH)_2$ is found in what medicine?

PRACTICE TEST II

1. What is the formula of the nitride ion?
2. What is the formula of the potassium ion?
3. Combine the two ions above into a compound and write its formula.
4. Write the formula for calcium iodide.
5. Write the formula for sodium bicarbonate.
6. Write the formula for stannic nitrite.
7. Name the compound Cu_3PO_3 using the old "-ous -ic" system.
8. Name the compound $Fe(BrO)_3$ using the Stock system.
9. Name the compound $AuIO_3$ using the Stock system.
10. Name the compound $Sn(CrO_4)_2$ by
    a) the Stock system
    b) the old "-ous -ic" system
11. Name the acid H_2CO_3 .
12. Name the acid H_2S .
13. Write the formula for acetic acid.
14. Name $MgSO_4{\cdot}7H_2O$.
15. Write the formula of chromium (III) nitrate nonahydrate, deep violet crystals.
16. Which of the following is used as a water conditioner?
    Na_2CO_3 H_2SO_4 $Ca(OH)_2$ N_2O
17. Ethanol, C_2H_5OH , is used
    A) in beer, wine, and liquor.
    B) as an antiseptic.
    C) in automotive fuel.
    D) in cologne.
    E) in all of the above.

Answers to Practice Test I

1. boride
2. Zn^{2+}
3. Zn_3B_2
4. Al_2S_3
5. $Zn(NO_3)_2$
6. $MgCl_2$
7. Li_2SO_3
8. $Sn(OH)_2$
9. $Fe(ClO)_2$
10. ammonium acetate
11. potassium cyanide
12. plumbous periodate
13. ammonium dichromate
14. cobalt (II) carbonate
15. potassium permanganate
16. CO_2
17. IF_7
18. PI_3
19. dinitrogen pentoxide
20. sulfur hexafluoride
21. +3
22. +4
23. +3
24. hydrobromic acid
25. sulfuric acid
26. $HClO_2$
27. cobalt (II) chloride hexahydrate
28. $Na_2S_2O_3$
29. NaOH (sodium hydroxide)
30. milk of magnesia

Answers to Practice Test II

1. N^{3-}
2. K^+
3. K_3P
4. BeI_2
5. $NaHCO_3$
6. $Sn(NO_2)_4$
7. cuprous phosphite
8. iron (III) hypobromite (by analogy to hypochlorite, ClO^-)
9. gold (I) iodate

10a. tin (IV) chromate
10b. stannic chromate

11. carbonic acid
12. hydrosulfuric acid
13. $HC_2H_3O_2$
14. magnesium sulfate heptahydrate
15. $Cr(NO_3)_3 \cdot 9H_2O$
16. Na_2CO_3
17. E

CHAPTER 7

Periodic Properties of Elements

In the introduction to Chapter 4 we wrote that chemists of the early 1800s sorely needed a way to categorize all elements but exclude all compounds. Mendeleev gave chemists a systematic table of elements. Mendeleev's great inspiration was leaving "holes" in his table. Few would think of doing this. Mendeleev used the holes to predict the existence of and the properties of elements not yet discovered. Mendeleev was right. All the holes were filled later.

Mendeleev recognized the *periodicity* of elemental properties. Neither he nor others realized how important this periodicity was. Atoms were believed to be indivisible (Dalton), that they had no structure. But periodicity in properties can occur only if an internal structure, which itself repeats in some fashion, exists.

The element's box indicated a close relationship between elements, a correlation Correlation leads to law. Law leads to theory. With theory, we begin to understand how things work, instead of merely noting that they do.

SKILLS TO ENHANCE SUCCESS IN THIS CHAPTER

Learn to "read" the periodic table. It truly is a picture worth a thousand words or, better yet, a thousand relationships and concepts. We can relate all of chemistry to the periodic table. Chemists use the table extensively. When physicists determined the structure of the atom, they adopted the table also. Students can benefit greatly by using it, too. You should know what the major sections are and review how they correspond to electronic structure. The table will serve as a template or model for chemical reactivity, valence, bonding, molecular structure, and many other properties. If the table's trends are used to full advantage, there will be much less need to memorize.

CHAPTER 7 PROBLEMS

7.1a. Mendeleev noticed that both the physical and chemical properties of elements varied periodically. Mendeleev was a chemist. He and others were primarily searching for order in chemical properties. What would you consider the most important classification of elements that is shown by the periodic table? Consider especially your study of ionic compounds.
Answer: One of the most important classifications is the grouping of metals and nonmetals using the staircase line. Mendeleev's stroke of genius was to arrange the elements according to atomic weight, a physical property. In so doing, he also arranged them according to chemical properties.

7.1b. What block(s) of the periodic table are occupied by metals? What blocks by nonmetals?

7.2a. What do the group numbers of the periodic table tell us of the chemical properties of the elements? Distinguish metals from nonmetals on this basis.
Answer: The column number is the number of electrons in the outer energy level of the atom. These are the valence electrons, the number of which determines the chemical properties of the element. The number of valence electrons in metals is low. The number of valence electrons in nonmetals is high. Chemically, a metal is an element that tends to donate its valence electrons. A nonmetal tends to accept more electrons into its valence energy level. We commonly think of metals as strong, used as support members in our structures. But some metals are soft. They are still metals chemically. Each column contains a *group* or *family* of elements. Elements within a *family* have similar chemical properties.

7.2b. Where are the metalloids on the periodic table? Are these metals or nonmetals? In what block is the zigzag or "staircase" line?

7.3a. Considering only the representative elements, what ion would be the largest ion possible?
Answer: Looking quickly at the periodic table, one would say the astatine ion, an anion of charge −1. But notice that the atomic radius of astatine is not much greater than that of iodine, immediately above it in the table. Astatine has an atomic number of 85; iodine is number 53. The nuclear charge of astatine is 1.6 times that of iodine. This charge pulls electrons strongly inward, so At^- may be largest, but not by much.

7.3b. What ion would be the smallest ion possible? What would be the value of its charge?

7.4a. From the periodic table alone, determine which element has the lowest first ionization energy. Then consult a table of ionization energies to confirm your prediction, or explain the discrepancy.
Answer: Ionization energy decreases leftward and downward on the periodic table. The element with the lowest ionization energy would probably be francium, since it occupies the leftmost and lowest position in the periodic table. Now look at text Figure 7.11. The ionization energy of francium is <u>not</u> the lowest. Cesium is lowest. Why? Because francium has a larger nuclear charge. This nuclear charge attracts electrons strongly, increasing ionization energy. Francium is only slightly larger than cesium, immediately above it in the table, so the distance of the outer electrons being ionized is about the same for Fr and Cs. But the pull of the nucleus is greater in francium, and that's why francium's electrons are slightly harder to remove than cesium's. And that's why cesium's ionization energy is lower than francium's.

7.4b. From the periodic table alone, determine which element has the highest first ionization energy. Then consult a table of ionization energies to confirm your prediction, or explain the discrepancy.

7.5a. What trend in melting points can be seen in the periodic table? Consider only representative elements.
Answer: Those elements near the zig-zag line, the metalloids, have the highest melting points. Moving away from this area, the melting points decrease, very rapidly in the nonmetals to the right. Look at Figure 7.13 of your text. Fluorine, F, is farthest from the zigzag line, and thus has the lowest melting point of the nonmetals. Cesium, Cs, is the farthest of the metals, and has the lowest melting point of the metals.

7.5b. What physical fact would lead one to expect that elements in column VIIIA have very low melting points?

7.6a. Explain the electrical conductivity of metals.
Answer: Metals tend to lose electrons from their valence sublevel. These electrons become "free" in the metal. Electron movement constitutes the electric current.

7.6b. Explain the electrical insulating characteristics of nonmetals.

7.7a. Explain the formation of ions by metals. Are these ions negatively or positively charged? What are these ions called?
Answer: Metals tend to lose their valence electrons during bond formation. A metal ion is positively charged, being formed by loss of negatively charged electrons. Positive ions are called *cations*.

7.7b. Explain the formation of ions by nonmetals. What is the sign on these ions? What are these ions called?

7.8a. Where in the transition elements would you expect to find the most electrically conductive elements?
Answer: In groups IB and IIB. Refer to Section 7.6 of your text.
Notice the column numbering in the transition block. Groups IB and IIB are on the right, not on the left as in the representative element blocks. This numbering implies that the elements in these groups hold at least some of their valence electrons more loosely than other transition elements.
The *d* orbitals are in use in the transition elements. The *d* orbitals form one energy level down from the *s* orbital of the same period. Even though these *d* orbitals are not in the outermost energy level, they are very close together (See Figure 5.36 of your text) and close to the *s* orbital of the outermost energy level. When all of the *d* orbitals are almost filled, the electrons, probably the *s* ones, are more free to move.
These are called transition elements because conceptually they join the *s* and *p* blocks. After an *s* orbital of a period is filled, the *p* orbitals of the same period must wait until the *d* orbitals are filled. The elements formed by the filling of the *d* orbitals are in the same period as those formed by the filling of the *s* and *p* orbitals, but the *d* orbitals are one *principal* energy level down.

7.8b. What element is used in the filaments of incandescent light bulbs? Why is this element used?

7.9a. What element should be the next one discovered? Would it be an inner transition element? To what use could it be put?
Answer: The next element would be element 110, a transition element, not an inner transition one. "Discovered" in this sense means man-made. All trans-uranium elements are man-made. They are highly radioactive and not found in nature. They have very short lives, some extremely small fractions of a second. They have no direct use. Producing them gives scientists (physicists) more knowledge of our universe. Production of these elements in accelerators has improved the theories of electronic structure and even slightly but importantly rearranged the periodic table. Read the caption of Figure 7.3 in your text.

7.9b. Into what block or blocks of the periodic table do the inner transition elements fit? What electron energy sublevels are occupied?

Answers to "b" Problems

7.1b. All elements in blocks *s*, *f* and *d* are metals. Some at the lower left of block *p*, to the left of and under the zigzag line, are metals. Nonmetals occupy block *p*, to the right of and above the zigzag line.

7.2b. The metalloids are adjacent to the zigzag line, which is within the *p* block. These elements have properties of both metals and nonmetals. The study of metalloids and the inclusion of the zigzag line on periodic tables is a development of the middle of this century.

7.3b. The hydrogen cation, of charge +1, would be the smallest possible. For normal hydrogen, this cation is nothing more than a proton.

7.4b. Helium would have the highest first ionization energy, according to the trend alone. It is a small atom with only two electrons in the *s* sublevel. These electrons are close to the nucleus and tightly held to the helium atom. From a table of ionization energies, the prediction turns out to be true.

7.5b. The column VIIIA elements do neither. In Chapter 5 you learned that these elements are the noble gases that have a complete octet of electrons, except for helium which has a complete shell 1 set of two electrons. Noble gases are all gaseous at room temperature; thus their boiling point is well below room temperature of 25°C, and their melting point is even lower.

7.6b. Nonmetals hold tightly to their valence electrons. Few electrons are free to conduct the electric current.

7.7b. Nonmetals tend to gain valence electrons during bond formation. The ion formed, called an anion, is negatively charged.

7.8b. Tungsten. With its high melting point (and boiling point), it can reach white heat and still retain its structural strength.

7.9b. The inner transition elements fit into the *f* block. The *s*, *p* and *d* and *f* sublevels are occupied.

PRACTICE TEST I

1. Who was the Russian that published the first periodic table that resembles the one we use today?
2. The metals of Group IA belong to what chemical family?
3. Give the chemical formula of any common halogen as it occurs in nature.
4. Which of the following subsets of elements occurs only in rows 4 through 7 of the periodic table in its current form?

   A) representative metals
   B) representative nonmetals
   C) metalloids
   D) transition metals
   E) halogens

5. Which of the following atoms is the largest in size, as measured by atomic radius? (Use only the periodic trend.)

   Na Li Rb K Cs

6. Compare the approximate relative size of the sodium atom, Na , with its cation, Na^+
7. Of the following, which element has the highest ionization energy? (Use only the periodic trend.)

   N Al F Br K

8. Which of the following Group IV elements has the highest melting point? (Use only the periodic trend.)

   carbon silicon tin lead

9. What chemical family on the periodic table is, on the whole, least reactive?
10. What group contains elements that are found in potato chips, computer chips, and paint chips?
11. Alloys of iron and carbon are known as steel. What other class of elements is often mixed with steel to give it special properties?
12. Describe where the inner transition metals lie in the periodic table.
13. Which is smallest?

    B C N O F

PRACTICE TEST II

1. What name is given to the Group II family?
2. Give the symbol of a second period element which is not a representative element.
3. What type of element is found to the upper right of the "staircase" line on the periodic table?
4. Which atom is the smallest, as measured by atomic radius? (Use only a periodic table.)

   He Ne Ar Kr Xe

5. Which atom is the largest, as measured by atomic radius? (Use only a periodic table.)

   Al Si P S Cl

6. The following ions are isoelectronic: Rb^{1+} Sr^{2+} Br^{1-} Se^{2-}. (All have 36 electrons.) Compare their relative sizes, using only a periodic table.

   A) $Se^{2-} < Br^{1-} < Rb^{1+} < Sr^{2+}$
   B) $Sr^{2+} < Rb^{1+} < Br^{1-} < Se^{2-}$
   C) $Br^{1-} < Se^{2-} < Rb^{1+} < Sr^{2+}$
   D) $Sr^{2+} < Se^{2-} < Rb^{1+} < Br^{1-}$
   E) $Rb^{1+} < Sr^{2+} < Br^{1-} < Se^{2-}$
   F) They are about the same size.

7. Which has the highest ionization energy, according to the periodic trend?

   Al F Mg S B

8. Which has the lowest melting point, according to the periodic trend?

   K Ca Sc Ti

9. Which has the highest density of copper, silver, and gold?

10. Which is the best conductor of electricity? Al Si P

11. Briefly describe hydrogen in terms of its

a) abundance, b) chemical reactivity, and c) chemical symbol and formula as it occurs in nature.

12. The alkali metals are never found in nature as pure elements because

A) they are too rare.
B) they are too reactive.
C) they are deep within the earth's crust.
D) they are relatively unreactive.
E) they are insoluble in water.

13. What metallic element is found in bones, eggshells, chalk, and limestone?

14. What charge is commonly held by Group I elements?

15. Who or what inspired the name for "Buckyballs", a C_{60} molecule shaped like a soccer ball?

A) George Villiers, Duke of Buckingham
B) American novelist Pearl S. Buck
C) Ernest "Buck" Rutherford, chemist
D) Buckminster Fuller, architect
E) a bucket of golf balls
F) a shoe buckle

16. What are the three members of the "coinage" family?

17. Noble gases are used

A) in medicines.
B) in electronic circuits.
C) inside light bulbs.
D) as fuels.
E) by royalty.

18. The elements that follow uranium in the periodic table are collectively known as transuranium elements. Circle all of the adjectives that apply to these elements.

metallic synthetic diatomic expensive naturally-occurring radioactive

Answers to Practice Test I

1. Mendeleev
2. alkali metals
3. F_2, Cl_2, Br_2, or I_2
4. D) transition metals
5. Cs
6. Na is about twice as large as Na^+.
7. F
8. carbon
9. Group VIII or the noble gases
10. Group IV (carbon, silicon, and lead)
11. transition metals
12. The two separate rows at the bottom contain the inner transition metals.
13. F

Answers to Practice Test II

1. alkaline earths (or alkaline earth metals)
2. There are none!
3. nonmetals (or representative nonmetals)
4. He
5. Al
6. B (The highest atomic number produces the most nuclear charge to rein in the electrons most closely.)
7. S
8. K
9. gold
10. Al
11. a) Hydrogen is the most abundant element in the universe.
    b) Hydrogen is explosive and flammable.
    c) Pure hydrogen occurs as H_2 in nature.
12. B
13. calcium
14. 1+
15. D
16. Cu Ag Au
17. C
18. metallic, synthetic, expensive, radioactive

CHAPTER 8

Chemical Bonds

Chemical bonds hold atoms together within a molecule. These connections result from interactions of *valence electrons*. The interactions are based on the properties of the atoms comprising the molecule. You will learn to predict the geometric structure of the molecules of a compound based on the orderly interaction of the atoms in the molecule.

SKILLS TO ENHANCE SUCCESS IN THIS CHAPTER

The ability to build mental images is essential to success in this chapter. You must be able not only to "see" a mental picture of a molecule but also to "play a movie clip" in your mind, watching atoms forming bonds during the building of the molecule.

You will extend to molecules the earlier-learned skill of reading and writing Lewis electron-dot formulas. These dot formulas or structures are pictures of electrons involved in bonding. Conscientiously practice the drawing method that you have developed. To be able to draw them shows that you are at least beginning to understand what the chemical bond is. Pictures help us to communicate with others. They also help us to communicate with ourselves; they aid us in our own thinking.

CHAPTER 8 PROBLEMS

We will present an alternate method of writing Lewis electron-dot formulas. Use this method only if it makes more sense to you than the method presented in the text.

8.1a. Given two binary compounds, both made up of a metal and a nonmetal, would you expect them to be equally ionic?
Answer: No two compounds are equally ionic, or identical in any property.
Our concept of the ionic bond is that of complete valence electron transfer from atoms of one element to atoms of the other. The concept, as powerful as it is in helping us understand the chemical bond, is nevertheless an ideal. Bonding is a spectrum, from extremely covalent to extremely ionic, with all the shades of gray in between. A bond could be moderately ionic, with some covalent character.
We have already learned that metals react with nonmetals. But some metals are more reactive than other metals, and some nonmetals are more reactive than other nonmetals.
Reading the periodic table, we know that the most reactive metals are on the lower left of the periodic table and that the most reactive nonmetals are on the top right of the periodic table. We neglect the noble gases, which have high ionization energies, because they are essentially unreactive.
On this basis we would expect that the most ionic binary compound would be formed between cesium and fluorine, that is, CsF.

8.1b. Which of these compounds would you judge to be the more ionic: lithium chloride or aluminum chloride?

8.2a. What constitutes the covalent bond?
Answer: A covalent bond consists of a pair of electrons under control of both atoms being bonded. *Co*valent means sharing of *valence* electrons. In the pure covalent bond, the two atoms equally control the electron pair. They are said to equally *share* the electron pair. This "50-50" perfectly equal sharing occurs only in a bond between two identical atoms. In compounds, one atom will have more control of the shared

electron pair than the other atom. The electron pair will be closer to the atom having more control. Sharing might be 60-40, or 90-10, depending on the difference between the two participants' electron "pull".

Note: Normally, one electron comes from each of the participants in a bond. However, to complete an octet, it is sometimes necessary that both shared electrons come from only one of the atoms. This is a special case called a coordinate covalent bond. Either way, once formed, the bond functions the same way.

8. 2b. In a nonpolar covalent bond, which atom has control of the electron pair of the bond? Give an example of such a bond. What is required of the two bonded atoms for a bond to be purely covalent?

8.3a. Electronegativity is described as a measure of the attraction that a covalently bonded atom has for electrons of the bond. Does electronegativity apply to ionically bonded atoms?

Answer: Yes. Electronegativity is the ultimate reason for both loss and gain of electrons prior to ionic bonding. An ionic bond is an extremely polar bond, in which one of the participants has a much stronger attraction for electrons than the other.

We can consider a "spectrum" of bond nature ranging from pure covalent to pure ionic. A pure covalent bond in which the electron pair is equally shared can occur only between atoms of the same element. As the atoms become increasingly dissimilar, covalency decreases and polarity increases. The bond becomes increasingly polar until it is finally called ionic.

The dissimilarity of the atoms in bond attraction is measured by electronegativity. The electronegativities of two elements, such as carbon and sulfur, may be reported as the same. Nevertheless, the elements are not identical. Electronegativities are difficult to measure, and are normally reported to only one decimal.

A difference of 1.7 in electronegativity between two elements is considered the crossover point.

Look at lithium chloride and aluminum chloride of the first problem. From Figure 8.6 of your text:

Electronegativity difference: Chlorine – Lithium = 3.0 – 1.0 = 2.0

Chlorine – Aluminum = 3.0 – 1.5 = 1.5

The bond between lithium and chlorine is considered ionic. The bond between aluminum and chlorine is covalent – highly polar.

8.3b. In the first problem, we expected that the most ionic binary compound would be formed between cesium and fluorine. On the basis of electronegativity, what compound would you predict to be the most ionic?

8.4a. In the compound carbon dioxide, which element functions as the metal? Which as the nonmetal?

Answer: Carbon functions as a metal, oxygen as a nonmetal, in carbon dioxide. The electronegativity difference between C and O is 1.0, with oxygen the more electronegative. Bonding is covalent in the CO_2 molecule. No ions exist in the solid. Both C and O are considered nonmetals. The question does not seem to make sense. Although we were introduced to bonding in simple terms of metals and/or nonmetals joining, the nature of the bond is not so "black and white". Ionic and covalent character is more of a continuum with every shade of gray in between. We can consider a bonding spectrum to exist on the basis of electronegativity difference.

The electronegativity of metals is low, and the electronegativity of nonmetals is high. As we move from left to right, electronegativity increases. As we move up a column, electronegativity increases. These observations lead us to considering degrees of metallic character corresponding to electronegativity value. As we move across the table from left to right, metallic character of the elements decreases. As we move up a group, metallic character of the elements decreases. The most metallic of the elements is the element at the bottom of group IA, leftmost and lowest. The most *un*metallic, i.e., nonmetallic, of the elements is the element at the top of group VIIA, rightmost and highest. As usual, we do not consider the noble gases of group VIIIA. In the sense that carbon is less electronegative than oxygen, it can also be deemed "more metallic" than oxygen. In carbon dioxide, oxygen has more control over the electron pair than does carbon, although carbon has not totally relinquished control of the electron pair as metals do. The C–O bond is polar. Both C and O atoms have *partial charges*, the C positive and the O negative.

Note: The great majority of elements are metals. Only the elements to the right of the zigzag line are nonmetals. In this course we have added the metalloids as a classification, but the metalloids overlap into the metals on the left and the nonmetals on the right of the periodic table. Elements close to the line can have "crossover" properties. For example, carbon does have some metallic physical properties. It has a

shiny surface reminiscent of the surface of metals. Graphite, the "lead" in pencils, is a conductor of electricity, and is used in the spark plug wires of your car's engine.
We consider aluminum to be a metal. It conducts like a metal, looks like a metal, and pounds flat like a metal. But aluminum can form covalent bonds as well as ionic ones. Its covalent bonds are polar. Most of its compounds are not very soluble in water, indicating weak ionic bonding. These effects are consistent with its position on the periodic table, adjacent to the zigzag line.

8.4b. In a compound of nitrogen and oxygen, which element functions as the metal, which as the nonmetal? The electron pair of the N–O bond is closer to which atom?

8.5a. Metals conduct both electricity and thermal energy ("heat") more easily than nonmetals, have reflective surfaces, and are opaque. Are all of these properties caused by the metallic bond?
Answer: Yes, all the properties are caused by the bonding within the mass of a metal. In a metal, electrons from outermost orbitals are free to move far away from their sources within the atoms. The movement of these electrons constitutes the electric "current" within the metal, hence conductivity.
The movement of electrons also greatly aids conduction of heat. Thermal energy is moved through a substance by the particles of the substance essentially jostling each other. The freer the particles are to move, the more rapidly they transfer the energy.
The large number of electrons on the surface of a metal reflect light. We see many metals as having a dull surface, but this is caused by an accumulation of a compound, usually an oxide, on the surface. A freshly cut surface of any metal will be shiny. The metal is opaque because the light is reflected. Even very thin films of metal are opaque.

8.5b. Metals are ductile and malleable. How are these properties caused by the metallic bond?

8.6a. A compound is highly soluble in water. Another compound is insoluble in water. Can you conclude that the first compound is ionic and the other molecular?
Answer: No. That a compound is soluble in water does not prove that it is ionic. You already know that ethanol and some other alcohols are miscible with water. The alcohols are molecular compounds. They dissolve in water by hydrogen bonding. Polarity of the molecules of these alcohols also facilitates dissolution in water. A molecular compound may dissolve in water by reacting with the water. Acids such as HCl and H_2SO_4, molecular compounds, react with water.
That a compound is insoluble in water does not prove that it's <u>not</u> ionic. Dissolution requires that the bonds holding the compound together be broken, and new bonds form between the solvent molecules and the molecules or ions of the compound. As long as the dissolving energy is sufficient to break the ionic bonds, the compound will dissolve. However, for many compounds of heavy metals, this is not the case. A number of ionic compounds are insoluble, such as silver chloride and lead sulfide.

8.6b. How would you go about showing that one of the compounds is ionic?

8.7a. Write the Lewis electron-dot formula for beryllium.
Answer:
Given: Symbolically represent one atom of Be.
Need: Number of valence electrons to draw Lewis electron-dot formula
Connecting Information:
Be is in Group II.

1. Draw the chemical symbol for beryllium.

Be

2. Draw in the two dots representing valence electrons, spreading them so that they are not on the same side of the beryllium atom.

• Be •

8.7b. Write the Lewis electron-dot formula for an atom of bromine.

8.8a. Write the Lewis electron-dot formula for elemental nitrogen.

Answer:

Given: Nitrogen occurs at normal conditions as a diatomic gas, N_2 molecules.

Need: Lewis electron-dot formula

Connecting Information:

This is a problem in construction of the Lewis formula for a molecule with multiply bonded atoms. The N–N bond is covalent and nonpolar, since two like atoms are joined.

1. Draw the Lewis electron-dot formula for a nitrogen atom.

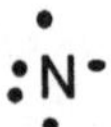

Consider one of the nitrogen atoms as "central".

2. Determine the minimum number of single bonds needed to connect the two nitrogen atoms.
At least one bond is needed to connect two atoms together. One nitrogen atom is joined to the other.

3. Join the two nitrogen atoms together by forming the first bond between the nitrogen atoms.
Reposition dots to prepare for bonding.

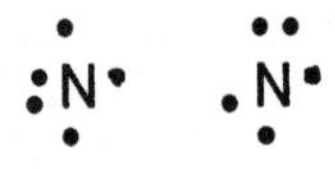

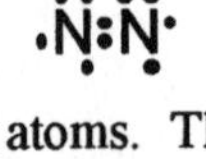

·N:N·

Note the unpaired electrons around adjacent atoms. These are to be paired to form multiple bonds.
We must make more bonds, because an octet is not yet present.
Join two more of the dots representing unpaired electrons to form a second bond between the nitrogen atoms.

·N::N·

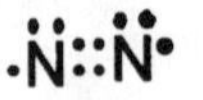

Join the other pair of electron dots to indicate a third bond between the nitrogen atoms.

N:::N

4. Construct the final formula.
Rearrange unbonded electron pairs on opposite sides to indicate maximum repulsion of electrons.

:N:::N:

5. Check the resulting bonding of each atom for either the noble gas octet arrangement of electrons or a special arrangement explainable in terms of the properties of the compound.

Nitrogen atoms: Each is surrounded by an octet. Bonding pairs are arranged symmetrically above and below the internuclear axis (an imaginary line drawn between the two N's).

:N :::N: :N ::: N:

The bonds are nonpolar, and the molecule is symmetric. Nitrogen gas is nonpolar.

8.8b. Write the Lewis electron-dot formula for CO_2.

8.9a. Write the Lewis electron-dot formula for the hypochlorite ion.

Given: ClO^-

Need: Lewis electron-dot formula.

Connecting Information:

This is a problem in construction of a Lewis electron-dot formula for a polyatomic ion.

1. Draw the Lewis electron-dot formulas for the atoms of the ion.

Draw the Lewis electron-dot formula for a chlorine atom.

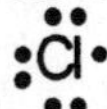

Draw the Lewis electron-dot formula for an oxygen atom.

2. Determine the number of single bonds needed to connect atoms in the ion: At least one bond is needed.

3. Indicate bonding of atoms.

Position the atomic symbols near each other.

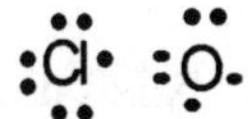

Note the position of unfilled orbitals. Reposition single dots in each atom to prepare for bonding. Join the atoms.

The arrow indicates a location still needing an electron. This is called a "hole".
One electron from an external source is present in the ion.
It is shown as a negative sign on the ClO^- chemical formula.
Add a dot indicating the extra electron, wherever it seems convenient – that is, in the hole.

$$:\ddot{\underset{\cdot\cdot}{Cl}}:\ddot{\underset{\cdot\cdot}{O}}:$$

4. Construct the final electron-dot formula.

The extra electron causes a negative charge on the structure. Always put big brackets around electron-dot structures that have a net charge, and write the charge outside the bracket on the upper right.

$$\left[:\ddot{\underset{\cdot\cdot}{Cl}}:\ddot{\underset{\cdot\cdot}{O}}:\right]^{1-}$$

5. Check the resulting structure of each atom for either the noble gas octet arrangement of electrons or an arrangement explainable in terms of the properties of the compound.

chlorine atom: Electrons are arranged around the atom in an octet.
oxygen atom: Electrons are arranged around the atom in an octet.
The bond of this ion is slightly polar.
It does, however, have a charge that is evenly distributed in three-dimensional space.

8.9b. Your text constructs the nitrate ion using resonance structures. Under what conditions should we repesent bonding of a compound or complex ion as a resonance structures? What do these structures say about the bonding? Hint: Think about the bonding of identical surrounding atoms to a central atom and the positioning of the electron pairs in all bonds about the central atom.

8.10a Considering geometry alone, what shape would you expect of a molecule consisting of four atoms bonded to a central atom?

Answer: A tetrahedral shape would be expected. Without additional information, one would assume a regular, symmetrical figure. Consult the drawing of the tetrahedral arrangement in section 8.18 of your text.

8.10b. What shape would you expect of a molecule consisting of two atoms bonded to a central atom?

8.11a Water is a planar molecule. Three points determine a plane. The three "points" are the centers of the three atoms forming a water molecule. Why then, to explain the shape of the water molecule, do we visualize the molecule as being within a tetrahedron, a solid figure?

Answer: That the water molecule is bent has long been known because the molecule exhibits polarity. If the molecule were linear, H–O–H, symmetry would cancel any internal polarity of each O–H bond. Original explanations of the bent shape involved repulsion of the two hydrogen nuclei, protons. As simple as the concept of electron pair repulsion is, it has not been applied for very many years. We know the concept as the VSEPR theory.

The oxygen is considered a *central* atom within the molecule. The oxygen atom is surrounded by four electron pairs. These pairs repel each other. They would be farthest from each other around a sphere, a three dimensional object, when they formed the legs and column of a symmetrical tripod, the outline of which is a regular tetrahedron. The hydrogens are not even considered in explaining the molecular shape.

8.11b. If the angle of the water molecule were changed, would this affect the physical properties such as viscosity and boiling point of liquid water? Hint: Think of hydrogen bonding in the water mass.

8.12a Your text describes ammonia as being a pyramidal molecule. Some molecules with three atoms surrounding a central atom have a flat trigonal shape in which all atoms lie in the same plane. How can we determine whether such a molecule is two or three dimensional?

Answer: We look not at the atoms surrounding the central atom but at the electron pairs surrounding the central atom. In the ammonia molecule, nitrogen is bonded to three hydrogens. Nitrogen has five valence electrons, hydrogen one each, for a total of eight. Four electron pairs, one unbonded, surround the central nitrogen atom. These four electron pairs are tetrahedral. The hydrogen atoms themselves form a pyramid. In boron trichloride, for example, the molecule is planar. Boron has three valence electrons. Chlorine has seven but only one chlorine electron is involved in bonding. Each chlorine contributes one electron to a bond with boron. The boron is the central atom, surrounded by six electrons, three pairs that mutually repel each other. They are flat, in a trigonal planar shape.

8.12b. The shape of the ammonia molecule is often described as a tripod. Is this a proper description of its shape?

8.13a For what reason is ammonium phosphate commonly used as a fertilizer?

Answer: If you're a gardener or a farmer, you know this one. Three elements necessary for plant growth are nitrogen, phosphoru,s and potassium. Ammonium phosphate contains both nitrogen and phosphorus.

8.13b. What are the shapes of the ammonium and phosphate ions? What are their bond angles?

8.14a Methane is a compound of carbon and hydrogen. Is methane soluble in water by hydrogen bonding?

Answer: No. For hydrogen bonding to occur between molecules, hydrogen must be bonded to a highly electronegative central atom such as oxygen, nitrogen, or fluorine. Methane has no hydrogen bonding.

8.14b. The electronegative difference between hydrogen and carbon in methane is 0.3, with carbon having the higher value. Is this small difference the reason that the methane molecule is nonpolar?

8.15a What is the shape of the molecule of phosphorus pentachloride?

Answer: Phosphorus has five valence electrons. With five chlorine atoms arranged around a phosphorus atom, five electron orbitals exist, all bonded to identical atoms. The molecule will be shaped to obtain as much symmetry as possible in three-dimensional space.

The shape of the molecule is described as trigonal pyramid. Three chlorines are symmetrically arranged around the central phosphorus atom in a triangular planar formation. The other two chlorines are above and below the plane of this triangle, forming the two joined pyramids. Think of the molecule as two pyramids with triangular bases joined base to base.

Note: Research indicates that this compound is actually P_2Cl_{10}, not PCl_5.

8.15b. A compound of oxygen and fluorine exists. Give the formula for this compound and give its name.

8.16a. What is required for hydrogen bonding to occur?
Answer: These conditions must prevail:
1. The molecule must be covalent, and contain hydrogen, of course!
2. The central atom must be highly electronegative, for example, N, O, or F.
3. The central atom must also be small (also N, O, or F).
4. The central atom must have lone (nonbonding) pairs of electrons.

8.16b. Water expands upon freezing because the water molecules do not "pack" in a dense crystalline structure. Why don't they?

Answers to "b" Problems

8.1b. Lithium chloride. Lithium, an alkali metal, is more reactive than aluminum. The difference in first ionization energy between lithium and aluminum is not greeat, but the ionization energy of lithium is lower. Further, aluminum is adjacent to the zigzag line and is one of the metalloids.

8.2b. Neither. The electron pair is equally shared. Examples: Diatomic element molecules: O_2, N_2, etc. For the bond to be totally nonpolar, the two atoms must be identical.

8.3b. Francium fluoride. But note that the differences in electronegativity between the alkali metals are small. We cannot easily distinguish the difference between KCl, RbCl, CsCl, or FrCl; all are highly ionic.

8.4b. N functions as the metal, O as the nonmetal. The electron pair is closer to the more electronegative oxygen.

8.5b. The metal atoms are not rigidly bonded to one another. The lattice is not fixed. The atoms can slide over one another without disturbing the metallic bond. Thus metals can be reshaped into wires or pounded flat.

8.6b. Check the indicative physical properties. A compound that is liquid or gaseous at standard temperature cannot be ionic. Ionic compounds conduct when melted or dissolved.

8.7b. $\cdot\ddot{\underset{\cdot\cdot}{\mathrm{Br}}}:$

8.8b. $:\underset{\cdot\cdot}{\mathrm{O}}::\mathrm{C}::\ddot{\mathrm{O}}:$

8.9b. Resonance structures are used where a central atom is bonded to two or more identical atoms but a double bond from the central atom to one or more of the surrounding atoms is required to complete the octet of the molecule or ion. No one surrounding atom can be bound differently to the central atom. Surrounding atoms, if all identical, must be bound identically to the central atom, and evenly spaced.

8.10b. Either linear or bent. Either way, the molecule would be planar. Three points determine a plane.

8.11b. Yes. Any change in the angle would change the polarity of the molecule and change the attraction of water molecules for one another. The bent shape of the water molecule allows the molecules of water to be virtually randomly distributed while still bonded to each other within the water mass.

8.12b. Both pyramid and tripod describe the same shape. *Tripod* emphasizes the bonded atoms of the molecule, with the bonds forming the "legs" of the tripod. The *pyramid* shape is an outline view of the outer faces, formed by the atoms at the corners of the shape. The bonds are within the pyramid.

8.13b. Both are pyramidal with angles of 109.5°.

8.14b. An electronegativity difference of 0.3 means that the individual bond between each C and H is classified as covalent. The bonds are very weakly polar. The entire CH_4 molecule (the sum of all four bonds) is nonpolar because the hydrogens are symmetrically distributed in space around the central carbon atom.

8.15b. OF_2. Oxygen difluoride. Fluorine has higher electronegativity and is treated as the negative side of the molecule. The name difluorine oxide would indicate that oxygen is negative.

8.16b. The bond angle of the water molecule is 104°, not a perfect tetrahedral angle of 109.5°. Its molecules do not stack up densely in the solid structure. Symmetry would occur if the water molecule angle were the ideal tetrahedral shape, and molecules would pack more compactly.

PRACTICE TEST I

1. What kind of bonding occurs in the compound Na_2S?
2. What type(s) of bonding is (are) described as fixed nuclei surrounded by a "sea" of loose electrons?
3. Liquor contains ethanol, C_2H_5OH. What type of bonding does it exhibit?
4. Draw the Lewis electron-dot structure for the calcium ion, showing the appropriate charge.
5. What type of bonding results in separate, discrete molecules?
6. Give an example of a real molecule that contains only one nonpolar bond.
7. Name the element with the highest electronegativity of all.
8. Arrange these in order of increasing electronegativity, using only the periodic table and the trends.

   Fr Al B Cl Mg

9. Which bond is most polar in character? F—F C—F H—C Al—Cl
10. Which is an electrolyte? CF_4 NO I_2 LiCl
11. Which has the highest melting point?

    carbon dioxide, CO_2 table salt, NaCl oil, $C_{16}H_{34}$
    table sugar, $C_{12}H_{22}O_{11}$ water, H_2O bromine, Br_2

12. An unidentified solid did not conduct electricity. After being melted, however, it did conduct electricity. What kind of bonding is present in the solid?
13. What is the molecular shape and polarity of CO_2? (Hint: After drawing the Lewis electron-dot structure, sketch in dipoles on top of the bonds and assess the total symmetry of the molecule. Is there any one direction in which the dipoles "gang up", or are they evenly distributed and opposed?)
14. What is the molecular shape of the molecule H_2O?
15. What are the molecular shapes of the following?

    CH_4 NH_3 BF_3 NH_4^+ CS_2

16. Which is trigonal planar?

    PCl_3 AsF_3 O_3 NI_3 CH_3^+

17. What Group often makes four bonds resulting in a tetrahedral shape?
18. Give an example of a substance that exhibits hydrogen bonding.
19. Draw the Lewis electron-dot structure for the H_2CO molecule.
20. Draw the Lewis electron-dot structure for the NO^+ (nitrosyl) ion.
21. Draw in the bonds and dipoles for this molecule: **N C H** (Lone pairs are not shown.)

    Is the molecule polar or nonpolar?

PRACTICE TEST II

1. What type of bonding occurs when one atom transfers electrons completely to another?
2. Electron sharing occurs in which of the following? (Circle answers.)

   $CaCl_2$ CH_4 F_2 NaK SF_6

3. A certain solid compound conducted electricity both as a solid and when melted. What kind of bonding is present in the solid?
4. Simple table sugar is sucrose, $C_{12}H_{22}O_{11}$. Describe its conductivity as a solid, when melted, and when dissolved in water.
5. Draw the Lewis electron-dot formula of the sodium ion.
6. Draw the Lewis electron-dot structure of sulfur monoxide, SO .
7. Draw the Lewis electron-dot structure of the cyanide ion, CN^- .
8. What is the shape of the SCl_2 molecule?
9. What is the shape of the CH_4 molecule?
10. Draw the Lewis electron-dot structure of the NOCl molecule. Careful selection of the central atom (which needs to make the most bonds) will result in all atoms having an octet by "fair" sharing of electrons, i.e. with no coordinate covalent bonds.
11. What is the shape of the C_2H_2 molecule?
12. A mystery atom makes a pyramidal molecule of formula XCl_3 . What Group is element "X" from?
13. Which of the following exhibit(s) 120° bond angles?

    CF_4 H_2O BH_3 NH_3 CO_2

14. Water exhibits unusual properties due to what type of bonding between two different molecules of water?
15. Determine the shape and overall polarity of the water molecule by assessing its shape and symmetry.
16. Draw in the dipoles for the bonds of the CO_2 molecule. Is the molecule polar or nonpolar?

Answers to Practice Test I

1. ionic
2. metallic
3. covalent
4. Ca^{2+} is the ion's formula. (All the valence electrons are gone.) $[Ca]^{2+}$
5. covalent
6. Any diatomic element: H_2, N_2, O_2, F_2, Cl_2, Br_2, or I_2.
7. fluorine
8. Fr < Mg < Al < B < Cl
9. Al — Cl
10. LiCl
11. NaCl (ionic)
12. ionic
13. linear, nonpolar molecule
14. bent
15. tetrahedral, trigonal pyramidal, trigonal planar, tetrahedral, linear
16. CH_3^+
17. Group IV
18. The compounds NH_3, H_2O, and HF are the three best examples.
19. H_2CO **H : C : : O :** (with **H** below C; electron dots ·· under C and O)
20. NO^+ **[: N : : Ö :]$^+$**
21. **N ←+ C ←+ H**

Answers to Practice Test II

1. ionic
2. CH_4, F_2, and SF_6
3. metallic
4. Sucrose, a covalent compound, does not conduct electricity under any of these conditions.
5. Na^+ is the ion. **[Na]$^+$** (All the electrons are gone.)
6. SO **: S̈ : : Ö :**
7. CN^- **[: C : : : N :]$^-$**
8. bent
9. tetrahedral
10. NOCl **: C̤̈l : N̈ : : Ö :**
11. linear
12. Group V
13. BH_3 only
14. hydrogen bonding
15. water is bent and polar
16. Carbon dioxide is nonpolar because the individual bond dipoles are symmetric and thus cancel out in this linear molecule.

O ←+ C +→ O

CHAPTER 9

Chemical Quantities

The concepts of this chapter connect the submicroscopic world of atoms and molecules to the *macroscopic* world, the world in which we live and perform measurements. Our symbolism represents simultaneously the submicroscopic world and the macroscopic world. Only through experimental verification can we know that the symbols we use do accurately represent both the submicroscopic and the macroscopic, and illustrate the relationships between the two worlds.

SKILLS TO ENHANCE SUCCESS IN THIS CHAPTER

You should learn how to move in the submicroscopic world and in the macroscopic world, and how to move from one world to the other. Before working a problem, determine to which world the problem primarily applies. The concept of the *mole* is a central concept in chemistry. The mole helps you move around in and between both worlds.

Avogadro's number defines the submicroscopic mole. *Molar mass* defines the macroscopic mole. You convert between mass and moles, between moles and number of molecules. The ability to use dimensional analysis to intelligently solve problems is essential. Let cancellation of units help you check your work.

CHAPTER 9 PROBLEMS

9.1a. Define and compare the following terms:

amu atomic mass atomic weight formula mass formula weight
molecular weight molecular mass molar mass mole

Answer: An **amu**, atomic mass unit, is an extremely small unit of mass usually used for measuring atomic mass. The present base for the atomic weight system is the isotope carbon-12, which is assigned the atomic mass of 12 amu. The unit itself was first defined by Dalton, in comparing the masses of other atoms to carbon's mass. We do not have balances that can actually weigh out amu's. **Atomic mass** or **atomic weight** is the mass of one atom of an element. The two terms are used interchangeably. The average atomic mass/weight can be found on most periodic tables at the bottom of each element's box. The units of atomic mass or atomic weight could be amu, if a single atom is being considered (microscopic scale), or grams per mole, if one mole (or any larger macroscopic amount) of the element is being considered.

The **formula mass** or **formula weight** of a substance is the sum of the atomic masses of all atoms in a substance as given by the formula of the substance. The terms are used interchangeably since mass and weight will have the same value – as long as we remain on planet Earth! Formula weight/mass is usually used for ionic compounds, and is calculated from the ratio of ions in the formula. For ionic compounds, it is more correct to use "formula weight" than "molecular weight", because ionic compounds have no "molecules". Even so, if you do say molecular weight when speaking of an ionic compound, everyone will still know what you mean, though it's technically incorrect.

The terms **molecular mass** and **molecular weight** are often used interchangeably. They are the mass of a molecular substance as calculated from its formula (the sum of individual weights of atoms in a molecule). **Molecular mass/weight** applies only to covalently bonded species that do exist as separate ("discrete") molecules. The units could be amu, if a single molecule is being considered, or grams per mole for one mole of the substance or more tangible amounts. The molecules could be either elements or compounds, but not single atoms.

The **molar mass** (sometimes called **molar weight**) of a substance is the mass of one mole (6.022×10^{23} particles) of the substance. It is calculated from the formula, and is always expressed in grams per mole. It can be used for atoms or molecules.

A **mole** is an amount of matter. A mole contains Avogadro's number of particles, where "particles" can mean atoms, ions, molecules, formula units, or even subatomic particles like electrons. Avogadro's number is 6.022×10^{23}. This is the number of particles which would be present in a sample whose mass is equal to the molar mass. If an ionic substance is being considered, a mole would be Avogadro's number of formula units.

9.1b. What is the average atomic mass in amu of nitrogen? What is its molar mass?

9.2a. Calculate the molar mass of thulium oxalate hexahydrate, $Tm_2(C_2O_4)_3{\cdot}6H_2O$.

Given: Atom moles: Tm: 2 C: 6 O: 18 H: 12 (see formula)

Need: $\frac{\text{grams}}{\text{mol}}$ of $Tm_2(C_2O_4)_3{\cdot}6H_2O$

Connecting Information: Atomic masses. Tm: 168.9 O: 16.00 C: 12.01 H: 1.008

Solution 1: Add the atomic masses directly, by element.

$$2(168.9) + 6(12.01) + 18(16.00) + 12(1.008) =$$
$$337.8 + 72.06 + 288.0 + 12.10 = 709.96$$

Consider significant figures.

$$710.0 \frac{\text{g}}{\text{mole}} \text{ of } Tm_2(C_2O_4)_3{\cdot}6H_2O$$

Solution 2: Calculate the molar masses of the actual atomic groupings and add their masses.

Calculate the molar masses of groups present within the molecule.

C_2O_4: $2 \times 12.01 + 4 \times 16.00 = 88.02 \frac{\text{g}}{\text{mol}}$

H_2O: $2 \times 1.008 + 1 \times 16.00 = 18.02 \frac{\text{g}}{\text{mol}}$

Calculate the total molar mass of the hydrate.

Mass of C_2O_4: $3 \times 88.02 = 264.06 \frac{\text{g } C_2O_4}{\text{mol hydrate}}$

Mass of H_2O: $6 \times 18.02 = 108.12 \frac{\text{g } H_2O}{\text{mol hydrate}}$

Mass of Tm: $2 \times 168.9 = 337.8 \frac{\text{g Tm}}{\text{mol hydrate}}$

Total: 709.98

$$710.0 \frac{\text{grams}}{\text{mol}} \text{ of } Tm_2(C_2O_4)_3{\cdot}6H_2O$$

You can use both methods to cross-check your calculations of molar mass.

9.2b. What is the molar mass of calcium nitrate monohydrate, $Ca(NO_3)_2.H_2O$?

9.3a. What is the percentage of water in the hydrate of problem 9.2a?

Given: Hydrate $Tm_2(C_2O_4)_3{\cdot}6H_2O$

Need: Mass percent of water in $Tm_2(C_2O_4)_3{\cdot}6H_2O$ (water's % mass out of the hydrate's total mass)

Connecting Information:

$\frac{710.0 \text{ g hydrate}}{\text{mol hydrate}}$ (Be sure to include six waters in your total.) 18.02 g/mole for water

Molar mass of six H_2O in hydrate: $6(18.02) = 108.1 \frac{\text{g } H_2O}{\text{mol hydrate}}$

Fraction of all H_2O in one mole of hydrate: $\frac{108.1 \text{ g } H_2O}{710.0 \text{ g hydrate}} \times 100\% = 15.23\%$

9.3b. What is the percentage of nitrogen in ammonium nitrate?

9.4a. How many moles of sodium chloride are there in 4.75 grams of the salt?
Given: 4.75 g NaCl
Need: moles NaCl

Connecting Information: $\frac{58.5 \text{ g NaCl}}{\text{mol NaCl}}$

$$4.75 \text{ g NaCl} \times \frac{1 \text{ mole NaCl}}{58.5 \text{ g NaCl}} = 8.12 \times 10^{-2} \text{ mole NaCl}$$

9.4b. What is the mass of 3.60×10^{-2} moles of $Tm_2(C_2O_4)_3 \cdot 6H_2O$?

9.5a. How many molecules of water are in 3.60×10^{-2} mole of thulium oxalate hexahydrate?
Given: 3.60×10^{-2} mole $Tm_2(C_2O_4)_3 \cdot 6H_2O$
Need: Number of water molecules
Connecting Information:

$\frac{6 \text{ mol } H_2O}{1 \text{ mol } Tm_2(C_2O_4)_3 \cdot 6H_2O}$

Avogadro's number: $6.02 \times 10^{23} \frac{\text{particles}}{\text{mol}}$

$$3.60 \times 10^{-2} \text{ mole } Tm_2(C_2O_4)_3 \cdot 6H_2O \times \frac{6 \text{ mol } H_2O}{1 \text{ mol } Tm_2(C_2O_4)_3 \cdot 6H_2O} \times \frac{6.02 \times 10^{23} \text{ molecules } H_2O}{\text{mole } H_2O}$$
$$= 1.30 \times 10^{23} \text{ molecules } H_2O$$

9.5b. How many atoms of oxygen are present in 1.75 moles of $Tm_2(C_2O_4)_3 \cdot 6H_2O$?

9.6a. You need 150. mL of a solution that is 1.50 M HCl and you have a solution that is 6.00 M HCl. How much of the concentrated solution do you need?
Given: Initial solution: 6.00 M
Final solution: 150. mL of 1.50 M
Need: mL initial solution
Connecting Information: The dilution equation is $M_{conc} V_{conc} = M_{dil} V_{dil}$
moles = MV (The number of moles of solute equals the molarity times the volume of the solution.)
Amount of HCl must stay constant: moles HCl in final solution = moles HCl in initial solution
Molarity ratio: $\frac{1.50}{6.00}$ (Final molarity over initial molarity)

Solution 1: Apply molarity ratio. $150 \text{ mL} \times \frac{1.50}{6.00} = 37.5 \text{ mL concentrated HCl}$

Solution 2: Apply dilution equation

$$M_{conc} V_{conc} = M_{dil} V_{dil}$$
$$6.00 \, M_{conc} \times V_{conc} = 1.50 \, M_{dil} \times 150 \, mL_{dil}$$

$$V_{conc} = \frac{1.5 \, M_{dil} \times 150 \, mL_{dil}}{6.0 \, M_{conc}} = 37.5 \text{ mL}$$

Solution 3: Apply equality of moles before and after. Moles in concentrated soln = moles in dilute soln.
Calculate moles of HCl in final dilute solution.
Moles HCl = 150 mL × 1.50 M = 225 mmole HCl
Calculate volume of initial concentrated solution.

$$\text{Volume concentrated HCl} = \frac{225 \text{mmolHCl}}{6.00 \text{M}} = 37.5 \text{ mL}$$

9.6b. How much concentrated (18.0 M) H_2SO_4 is needed to make a quarter liter of 6.00 M acid?

9.7a. A compound is composed of carbon and hydrogen. Its molecular weight is 30. What is the molecular formula for the compound? What is its empirical formula?
Given: Compound C_nH_m

Molecular weight of compound: 30

Need: n and m in molecular and empirical formulae

Connecting Information:

Atomic weights

Carbon: 12

Hydrogen: 1

Valences

Carbon: 4

Hydrogen: 1

1. Determine n, the number of carbon atoms in the molecule.

Three carbon atoms would contribute a value of 36 to the molecular weight.

With only one carbon atom, 18 hydrogen atoms would be present to add to the molecular weight.

Carbon has 4 valence electrons, hydrogen 1 valence electron.

The ratio of hydrogen atoms to carbon atoms in the molecule must be 4 or less.

$n = 2$

2. Determine the number of hydrogen atoms in the molecule.

With 2 carbon atoms, the number of hydrogen atoms must be 6:

Molecular weight $= 2 * 12 + 6 * 1 = 30$

3. Determine the molecular formula.

The ratio of hydrogen atoms to carbon atoms in the molecule:

Hydrogen atoms to carbon atoms $= 6/2 = 3$

The molecular formula is, then:

Molecular formula: C_2H_6

4. Determine the empirical formula.

The empirical formula is obtained by dividing both subscripts by 2:

Empirical formula: CH_3

9.7b. The empirical formula of a compound is HO. Its molecular weight is 34. What is the molecular formula of the compound?

9.8a. A common anti-inflammatory drug is 73.72% carbon, 5.12% hydrogen, 4.78% nitrogen, and 16.38% oxygen. What is the empirical formula of this drug?

Given: Compound: 73.70% C 5.16% H 4.78% N 16.36% O

Need: Values of subscripts i, j, k and l in empirical formula $C_iH_jN_kO_l$

Connecting Information:

Formula subscripts must be whole numbers.

Atomic masses:

C: 12.01 H: 1.01

N: 14.01 O: 16.00

1. Calculate the number of moles of each element in 100 g of the compound.

$$73.70 \text{ g C} \times \frac{1 \text{ mol C}}{12.01 \text{ g C}} = 6.14 \text{ mole C}$$

$$5.16 \text{ g H} \times \frac{1 \text{ mol H}}{1.01 \text{ g H}} = 5.11 \text{ mole H}$$

$$4.78 \text{ g N} \times \frac{1 \text{ mol N}}{14.01 \text{ g N}} = 0.341 \text{ mole N}$$

$$16.36 \text{ g O} \times \frac{1 \text{ mol O}}{16.00 \text{ g O}} = 1.02 \text{ mole O}$$

The smallest number is that of N.

2. Calculate the subscripts in the formula.

Assume that the subscript on N in the formula for the compound is unity (one).

N: 1

C: $\frac{6.14}{0.341} = 18.01$

H: $\frac{5.11}{0.341} = 14.98$

O: $\frac{1.20}{0.341} = 2.99$

3. Write the formula.

The calculated values of subscripts are close to whole numbers.

The empirical formula can be written:

$C_{18}H_{15}NO_3$

If the first cut would have resulted in the value of any subscript not close to a whole number, the subscripts would be adjusted by multiplying by 2, 3, 4 or higher integer. For example, if any value were closer to a half between whole numbers, all subscript values would be multiplied by 2.

9.8b. A compound is determined to be 15.8% carbon and 84.2% sulfur. What is its empirical formula?

9.9a. If the molar mass of the compound of problem 9.8b is 76 $\frac{g}{mol}$, what is the compound's molecular formula?

Given: Molar mass: 76 $\frac{g}{mol}$

Need: Values of subscripts i and j in molecular formula C_iS_j

Connecting Information:

Empirical formula: CS_2

Empirical mass: 76

1. Determine the empirical formula.

$$\text{Moles carbon in 100 g} = 15.8 \text{ g C} \times \frac{1 \text{ mol C}}{12.01 \text{ g C}} = 1.32 \text{ mole}$$

$$\text{Moles sulfur in 100 g} = 84.2 \text{ g S} \times \frac{1 \text{ mol S}}{32.06 \text{ g S}} = 2.63 \text{ mole}$$

Empirical formula: $C_{1.32}S_{2.63}$ or CS_2

2. Calculate the empirical formula mass.

Formula mass = atomic mass carbon + atomic mass sulfur

Formula mass = $1 \times 12 + 2 \times 32 = 12 + 64 = 76$

3. Determine the molecular formula.

Calculate the multiplying factor applied to each subscript in the empirical formula.

$$\text{Multiplying factor} = \frac{\text{molar mass}}{\text{empirical mass}} = 1$$

The molecular formula is, then, $\mathbf{CS_2}$

9.9b. The molar mass of the drug in problem 9.8a is 586 $\frac{g}{mol}$. What is the molecular formula for this compound?

Answers to "b" Problems

9.1b. The average atomic mass value is 14.0 amu (or 14.0 g/mole). The molar mass is 28.0 g/mole for N_2.

9.2b. $182.08 \frac{g\ Ca(NO_3)_2 \cdot H_2O}{mole\ Ca(NO_3)_2 \cdot H_2O}$

9.3b. 35.0% nitrogen

9.4b. 25.6 g

9.5b. 1.90×10^{25} atoms of oxygen

9.6b. 83.3 mL of 18.0 M H_2SO_4

9.7b. H_2O_2. Molecular weight = 2 * 1 + 2 * 16 = 2 + 32 = 34 g/mole.

9.8b. CS_2

9.9b. $C_{36}H_{30}N_2O_6$

PRACTICE TEST I

1. What is the molar mass (or molecular weight) of water?
2. What is the molecular weight of caffeine, $C_8H_{10}N_4O_2$?
3. What is the molecular weight of BHT, a food preservative, whose formula is $C_6H_2[(CH_3)_3C]_2OHCH_3$?
4. How many moles of hydrogen atoms are present in 0.125 moles of butylated hydroxytoluene (BHT)?
5. How many moles of oxygen atoms are present in 10.0 moles of washing soda, $Na_2CO_3 \cdot 10H_2O$?
6. How many grams are in 0.25 moles of BHT ($C_{15}H_{24}O$)?
7. How many moles are present in 2.0 tons of pure calcite limestone (calcium carbonate)?
8. How many atoms of copper are there in a 1.00 kg piece of copper pipe?
9. What is the mass in grams of 3.01×10^{22} molecules of water?
10. What is the mass in grams of 1 molecule of water? (See #9.)
11. A gallon of water has how many molecules? (density of water = 1.0 g/mL, 1 L = 1.06 quarts)
12. What is the molarity of a solution made by dissolving 2.0 moles of solute in enough water to make 500. mL of final solution?
13. What is the molarity of a solution made by dissolving 10.00 g of sodium hydroxide up to 800.0 mL of solution?
14. How many grams of glucose, $C_6H_{12}O_6$, are needed to make 500. mL of a 1.50 M solution?
15. You require 0.040 moles of potassium chloride. What volume of 0.200 M solution will provide this amount?
16. How many kilograms of sodium hydroxide are needed to prepare 100. L of a 3.0 M solution?
17. What is the final concentration of a solution made by taking 10.0 mL of a 5.0 M solution and diluting it to 250 mL?
18. What volume of 6.0 M NaOH is needed to prepare 48 L of a 0.50 M solution?
19. How you would prepare 500. mL of 0.500 M ammonia from concentrated stock solution of 12.0 M ammonia?
20. What is the percent by weight of oxygen in water?
21. What is the percent by weight of sodium in sodium sulfate, Na_2SO_4?
22. What is the percent by weight of oxygen in sodium carbonate decahydrate ($Na_2CO_3 \cdot 10H_2O$)?
23. What is the percent by weight of water in sodium carbonate decahydrate?
24. A compound is analyzed and is found to be 71.4 % calcium and 28.6 % oxygen. What is its empirical formula?
25. A compound is analyzed and found to be 81.81 % carbon and 18.18 % hydrogen. What is its empirical formula?
26. Elemental analysis of a salt provides the following data:

    Na 29.1 %

    S 40.5 %

    O 30.4 %

    What is the empirical formula of the salt?
27. A compound with a molecular weight of 32.0 g/mole was analyzed and found to be 87.5 % nitrogen and 12.5 % hydrogen. What is the molecular formula?
28. Ethyl maleate has a molecular weight of 172 g/mole and the following composition:

    C 55.8 %

    H 6.98 %

    O 37.2 %

    What is its molecular formula?

PRACTICE TEST II

1. What is the molecular weight of aluminum thiosulfate, $Al_2(S_2O_3)_3$?
2. What is the percent by weight of aluminum in aluminum thiosulfate?
3. What is the percent by weight of sulfur in aluminum thiosulfate?
4. What is the percent by weight of water in copper (II) sulfate pentahydrate?
5. What is the mass of 1.50 moles of sodium sulfide, Na_2S ?
6. How many moles of gold (Au) are present in one ounce? (16 oz/lb)
7. How many gold atoms are there in 4.925 g of gold?
8. How many moles of oxygen atoms are present in 0.25 moles of alum, $KAl(SO_4)_2 \cdot 12H_2O$?
9. How many grams of water are present in 0.25 moles of alum?
10. If you had 90.00 kg of scrap aluminum, how many moles of aluminum is that?
11. What mass of **alum** can be produced from 9.00 kg of scrap **aluminum**, if there is an ample supply of the other components? (The molecular weight of alum is 474 g/mole.) Hint: Use a mole ratio, just as you did in #8 and #9, only you are starting with the individual atom and relating it back to the compound.
12. What is the molarity of a solution made by dissolving 40.0 g of chromium (VI) oxide, CrO_3 , in enough water to make 200. mL of solution?
13. What mass of sucrose, $C_{12}H_{22}O_{11}$, is needed to make 2.50 L of a 2.00 M solution?
14. What volume of 0.456 M sodium hydroxide is needed to deliver 0.114 moles of solute?
15. What volume of 6.0 M stock hydrochloric acid is needed to make 100. mL of 1.5 M HCl ?
16. What is the empirical formula of a compound that is analyzed and found to be 15.8 % carbon and 84.2 % sulfur by weight?
17. What is the empirical formula of a compound that is 36.8 % nitrogen and 63.2 % oxygen by weight?
18. Vitamin C (ascorbic acid) has a molecular weight of 176 g/mole and is analyzed to contain:

C	40.9 %
H	4.54 %
O	54.5 %

What is the molecular formula of vitamin C?

Answers to Practice Test I

1. 18.0 g/mole
2. 194 g/mole
3. 220. g/mole
4. 3.00 moles
5. 130. moles
6. 55 g
7. 1.8×10^4 moles
8. 9.47×10^{24} atoms
9. 0.900 g
10. 2.99×10^{-23} g
11. 1.3×10^{26} molecules
12. 4.0 M
13. 0.3125 M
14. 135 g
15. 0.20 L, or 200 mL
16. 12 kg
17. 0.20 M
18. 4.0 L
19. Measure out 20.8 mL of 12.0 M ammonia and dilute it with water to a total volume of 500. mL, mixing well.
20. 89 %
21. 32 %
22. 73 %
23. 63 %
24. CaO
25. C_3H_8
26. $Na_2S_2O_3$
27. N_2H_4 (empirical formula NH_2)
28. $C_8H_{12}O_4$

Answers to Practice Test II

1. 390 g/mole
2. 14 %
3. 49 %
4. 36 %
5. 117 g
6. 0.144 moles
7. 1.505×10^{22} atoms
8. 5.0 moles
9. 54 g
10. 3333 moles
11. 158 kg
12. 2.00 M
13. 1710 g
14. 0.250 L
15. 0.025 L, or 25 mL
16. CS_2
17. N_2O_3
18. $C_6H_8O_6$

CHAPTER 10

Chemical Reactions

Any field has its terminology, verbal and written. The symbols for the elements and the formulae for compounds are the written shorthand of science.

The universe is driven by nuclear reactions, the reactions of the stars. Life is driven by chemical reactions. Each one of us is a chemical factory. We are chemical.

Physicists study nuclear reactions. Chemists study chemical reactio ns. Chemical reactions involve changes in matter and energy. Nuclear reactions involve interconversion of matter and energy. Chemists take the classical approach of separating the calculations of matter and the calculations of energy as if matter and energy had separate existences. This approach works because both matter and energy are each conserved in a chemical reaction.

SKILLS TO ENHANCE SUCCESS IN THIS CHAPTER

The symbols of elements and the formulas of compounds represent substances. Equations represent events: substances interacting with one another forming new substances.

Try to think in terms of proportion to best understand chemical reactions. A chemical equation does not tell us the quantities of substances reacting with one another or the quantities of products. It does tell us the proportional amounts of reactants required and products produced.

And, oh yes, do not forget moles. An equation says that for every mole of one reactant so many moles of another reactant will be required and so many moles of each product will be produced. Molar quantities are directly indicated by the *coefficients* of an equation. Mass in normal weight units have to be calculated.

You can use pounds or tons or any other weight unit in your mass calculations. We normally use grams.

CHAPTER 10 PROBLEMS

10. 1a What does a chemical equation "say" to us?

Answer: A chemical equation describes a reaction using symbolic language:

- Macroscopic
    - Identities of the chemical species involved
        - Reactants
        - Products
    - Which elements are making and/or breaking bonds to other elements.
    - The mole ratio (proportional amounts of each species involved)
    - The conditions of the reaction (limited; see below)
    - Energy effects (if included in the equation)
- Microscopic
    - Identity of particles involved; atoms, molecules, ions
    - Number of particles of each type and species involved
    - Which atoms are making and/or breaking bonds to other atoms
    - Which polyatomic groups remain unchanged during the reaction (spectator ions)

An equation can give us a clue to reaction conditions. Notations regarding state, energy, and catalysts help in this respect. An equation written in ionic form tells us that the reaction occurs between ions, whether in a melt or, more usually, in solution. The exact conditions of reaction are not given in detail by the symbolic equation.

This is all an individual equation can directly tell us. But if we look at many equations, which is to say look at many reactions, we can generalize to profound conclusions. The early chemists did just that. They discovered atoms and molecules this way. They discovered the laws of definite and multiple proportions.

10.1b. What cannot be told from an individual chemical equation? List a few factors.

10.2a. Determine whether the law of mass conservation applies to the reaction given.

Given: Equation of reaction: $Na_2CO_3 + 2\,HCl \longrightarrow 2\,NaCl + CO_2 + H_2O$

Need: Mass balance

Connecting Information:

Atomic masses: Na: 22.99 C: 12.01 O: 16.00 H: 1.008 Cl: 35.45

Molar mass = sum of atomic masses in formula

Coefficient multiplies molar mass.

Solution 1: Step-by-step

1. Calculate molar masses of reactants and products. "Mass units" can be either amus or g/mole.

$$Na_2CO_3:\quad 2(22.99) + 12.01 + 3(16.00) = 45.98 + 12.01 + 48.00 = 105.99\,\frac{\text{mass units}}{\text{mol}}$$

$$HCl:\quad 1.008 + 35.45 = 36.458\,\frac{\text{mass units}}{\text{mol}}$$

$$NaCl:\quad 22.99 + 35.45 = 58.44\,\frac{\text{mass units}}{\text{mol}}$$

$$CO_2:\quad 12.01 + 2 \times 16.00 = 44.01\,\frac{\text{mass units}}{\text{mol}}$$

$$H_2O:\quad 2 \times 1.008 + 16.00 = 18.016\,\frac{\text{mass units}}{\text{mol}}$$

2. Calculate total molar mass of reactants. Mass units remain all amus or all g/mole.

$$\text{Total reactant molar mass} = \text{Molar mass } Na_2CO_3 + 2 \times \text{Molar mass HCl}$$

$$105.99 + 2(36.458) = 178.906 \text{ mass units}$$

3. Calculate total molar mass of products.

$$\text{Total product molar mass} = 2 \times \text{molar mass NaCl} + \text{molar mass } CO_2 + \text{molar mass } H_2O$$

$$2 \times 58.44 + 44.01 + 18.016 = 178.906 \text{ mass units}$$

4. Equate the reactant molar masses to the product molar masses.

Reactant molar mass = 178.906 mass units

Product molar mass = 178.906 mass units

The law of conservation of mass applies.

The unit $\frac{\text{mass units}}{\text{mol}}$ is used to emphasize that a mole is a dimension, a unit of mass, which can be used with any practical mass unit. You would normally use the unit $\frac{g}{\text{mol}}$.

Solution 2: Proportions

Equation:	Na_2CO_3	+	$2\,HCl$	$\longrightarrow$	$2\,NaCl$	+	CO_2	+	H_2O
Molar mass:	105.99		36.458		58.44		44.01		18.016
Moles:	1		2		2		1		1
Equation mass:	105.99		72.916		116.88		44.01		18.016
Total equation mass:		178.906		=		178.906			

Equation mass = molar mass × moles

A tabular form presents the information in a concise way and makes the calculations easier to check. Label the lines as you see fit. You will be doing calculations about equations in the next chapter.

10.2b. Determine whether or not the law of mass conservation applies to this equation:

$$CH_4 + 2\,O_2 \longrightarrow CO_2 + 2\,H_2O$$

10.3a. Balance this equation representing the reaction producing oxygen during photosynthesis by plants.

$$CO_2 + H_2O \longrightarrow C_6H_{12}O_6 + O_2$$

Answer:

1. Look first at carbon.

Six "carbons" are shown on the right in the products.
Place coefficient 6 in front of CO_2.

$$6\ CO_2 + H_2O \longrightarrow C_6H_{12}O_6 + O_2$$

2. Look next at hydrogen.

Twelve "hydrogens" are shown on the right in the products.
Place coefficient 6 in front of H_2O.

$$6\ CO_2 + 6\ H_2O \longrightarrow C_6H_{12}O_6 + O_2$$

3. Oxygen remains unbalanced.

18 "oxygens" are shown in the reactants, 8 in the products.
Place coefficient 6 in front of O_2.

$$6\ CO_2 + 6\ H_2O \longrightarrow C_6H_{12}O_6 + 6\ O_2$$

4. Check equation balance by counting atoms.

Reactants		Products
6	C	6
$6 \times 2 = 12$	H	12
$6 \times 2 + 6 = 12 + 6 = 18$	O	$6 + 6 \times 2 = 6 + 12 = 18$

The equation is balanced.

10.3b. Balance this equation:

$$C_4H_{10} + O_2 \longrightarrow CO_2 + H_2O$$

10.4a. How would you classify a neutralization reaction?
Answer: A neutralization reaction is a reaction between an acid and a base that produces water and a salt. It is carried out in water solution. A neutralization reaction can be written in this general form:

$$HA + BOH \longrightarrow HOH + BA$$

This form shows a neutralization reaction to be a double-replacement, metathesis, reaction.
A neutralization reaction can also be classified as ionic. See problem 10.12a.
Note: A neutralization reaction is one that produces the solvent and a salt of some form. Solvents other than water can be used for certain reactions.

10.4b. Metallic zinc reacts with hydrochloric acid. How would you classify this reaction?

10.5a. Write, balance, and classify the reaction of propane and oxygen. (Propane is C_3H_8.)
Answer:

$$C_3H_8 + 5\ O_2 \longrightarrow 3\ CO_2 + 4\ H_2O$$

The reaction is a combustion reaction.

10.5b. Classify, complete, and balance the reaction of methanol and oxygen.

10.6a. Is this reaction a synthesis reaction?

$$P_4O_{10} + 6\ H_2O \longrightarrow 4\ H_3PO_4$$

Answer: The reaction is a synthesis reaction. It corresponds to the general formula:
$A + B \longrightarrow AB$ Only one product is formed.

10.6b. Write the synthesis reaction for the reaction between calcium oxide and carbon dioxide.

10.7a. Zinc-coated steel sheet metal has long been used as roofing and siding on light-duty commercial buildings and farm structures. What purpose does the zinc coating serve?
Answer: The zinc coating protects the iron from rusting – for a while at least.
Look at the Activity Series of Metals, Table 10.1, in your text.
A metal is more easily oxidized than a metal below it in the activity series. Zinc, above iron on this table, will replace iron in reactions. This replacement is shown by the net ionic equation:

$$Zn + Fe^{2+} \longrightarrow Zn^{2+} + Fe$$

The zinc corrodes preferentially to the iron.

10.7b. Our airplanes are made of aluminum because it is low in density. Would you suggest coating the skin of an airplane with sodium to minimize corrosion of the aluminum?

10.8a. Silver nitrate and sodium chloride will react, in water solution, in a double-replacement reaction. Write the equation for the reaction and show that it is a double-replacement reaction.
Answer: The reaction is:

$$AgNO_3 + NaCl \longrightarrow AgCl\downarrow + NaNO_3$$

The down arrow indicates that AgCl is a precipitate.
The reaction mass consists of ions:

$$Ag^+ \quad NO_3^- \quad Na^+ \quad Cl^-$$

The silver and sodium have switched partners. For the reaction to proceed to the right, producing the products shown, a product that escapes from the reacting mass must be formed. This product is AgCl, which is insoluble. As AgCl precipitates, the silver and chloride ions are removed from the reacting mass. The sodium and chlorine ions remain behind as a solution of ordinary salt, NaCl.

10.8b. How would potassium chromate, K_2CrO_4, react with lead(II) nitrate? Write the balanced equation for this reaction and classify it.

10.9a. Write the full ionic and net ionic equations for the neutralization of sulfuric acid by aluminum hydroxide. Which ions are spectator ions?
Answer: The full ionic equation is:

$$6H^+ + 3\,SO_4^{2-} + 2\,Al^{3+} + 6\,OH^- \longrightarrow 6\,HOH + 2\,Al^{3+} + 3\,SO_4^{2-}$$

The net ionic equation is:

$$6H^+ + 6\,OH^- \longrightarrow 6\,HOH$$

The coefficients 6 have been left on the ions for comparison with the full ionic equation. Reduced, the equation becomes:

$$H^+ + OH^- \longrightarrow HOH$$

This is the neutralization reaction in water solution. It is the bottom line, the important part of the reaction. All other ions are spectator ions, meaning they do not actually react, but only serve as the original partners of the reacting ions. The spectator ions are SO_4^{2-} and Al^{3+}.
The salt $Al_2(SO_4)_3$ remains dissolved at low concentrations. It appears as a solid if the resulting solution is evaporated. Although spectator ions do cancel out of the net equation, it doesn't mean they aren't there. They are still physically present. The net equation only shows the species that have actually reacted and changed, not those that are just "hanging around".

10.9b. Write the full ionic and net ionic equations for the neutralization of phosphoric acid by sodium hydroxide.

10.10a. What is the general neutralization reaction? What are the spectator ions in a neutralization reaction?
Answer: The general neutralization reaction is:

$$\text{Acid} + \text{Base} \longrightarrow \text{Water} + \text{Salt}$$

"Salt" is a generic term for an ionic compound. A salt is produced when an acid reacts with a base. Rewrite the general equation in ionic form.

$$H^+ + A^- \quad + \quad B^+ + OH^- \longrightarrow HOH \quad + \quad A^- + B^+$$

The spectator ions are the anion of the acid and the cation of the base.

$$A^- \quad B^+$$

10.10b. Write the neutralization reaction between acetic acid (in vinegar) and sodium bicarbonate (baking soda). Be sure to give the final products. Hint: What is baking soda used for?

Answers to "b" Problems

10.1b. The equation cannot tell us:
- Total amounts of species involved
- Speed (rate) of the reaction
- The complete set of conditions necessary for the reaction to occur
- How to experimentally identify the products
- Whether or not the reaction will occur, or to what extent it will occur
- Energy involvement (unless specifically included in the symbolism)

10.2b. Mass reactants = 16.042 + 64 = 80.042 = mass of products = 44.01 + 36.032 = 80.042

10.3b. $2\ C_4H_{10} + 13\ O_2 \longrightarrow 8\ CO_2 + 10\ H_2O$

10.4b. Single-replacement

10.5b. $2\ CH_3OH + 3\ O_2 \longrightarrow 2\ CO_2 + 4\ H_2O$ Combustion

10.6b. $CaO + CO_2 \longrightarrow CaCO_3$

10.7b. Sodium is an alkali metal that is quite reactive. It reacts violently with water, producing sodium hydroxide (lye) and explosive hydrogen gas. Sounds rather dangerous, don't you think?

10.8b. Both reactants are soluble in water and react in water solution.

$K_2CrO_4 + Pb(NO_3)_2 \longrightarrow PbCrO_4\downarrow + 2KNO_3$

The reaction is a double-replacement one. Lead (II) chromate is insoluble in water.

10.9b. Full ionic equation: $3\ H^+ + PO_4^{3-} + 3\ Na^+ + 3\ OH^- \longrightarrow 3\ H_2O + 3\ Na^+ + PO_4^{3-}$

Net Ionic equation: $3H^+ + 3OH^- \longrightarrow 3H_2O$ It's OK to leave out the threes.

Spectator ions: Na^+ PO_4^{3-}

10.10b. The baking soda is used as a *leavening* agent, which produces a gas. The gas is carbon dioxide.

The first equation to write is:

$HC_2H_3O_2 + NaHCO_3 \longrightarrow NaC_2H_3O_2 + H_2CO_3$

The carbonic acid decomposes to carbon dioxide and water:

$H_2CO_3 \longrightarrow CO_2\uparrow + HOH$

The carbon dioxide bubbles off. The overall reaction is, then:

$HC_2H_3O_2 + NaHCO_3 \longrightarrow NaC_2H_3O_2 + HOH + CO_2$

PRACTICE TEST I

1. Write a balanced chemical equation for the reaction of aqueous lead (II) perchlorate with a solution of potassium sulfate to form aqueous potassium perchlorate and a solid precipitate of lead (II) sulfate.

2. Write a balanced chemical equation for the reaction of vinegar and baking soda to form aqueous sodium acetate, water, and carbon dioxide.

3. Balance the following equation for the production of hydrogen bromide.

$$H_{2(g)} + Cl_{2(g)} \rightarrow HBr_{(g)}$$

4. Balance the following equation for an interhalogen reaction.

$$I_2 + F_2 \rightarrow IF_7$$

5. Balance the following equation for the precipitation of silver chloride.

$$AgNO_{3(aq)} + BaCl_{2(aq)} \rightarrow AgCl_{(s)} + Ba(NO_3)_{2(aq)}$$

6. Balance the following metathesis equation.

$$CaCl_{2(aq)} + (NH_4)_3PO_{4(aq)} \rightarrow Ca_3(PO_4)_{2(s)} + NH_4Cl_{(aq)}$$

7. Balance the following equation for iron production from iron ore.

$$Fe_2O_3 + CO \rightarrow Fe + CO_2$$

8. Balance the following equation for the explosion of nitroglycerin.

$C_3H_5N_3O_9 \rightarrow N_2 + CO_2 + H_2O + O_2$

CLASSIFY THE FOLLOWING REACTIONS AS COMBINATION, DECOMPOSITION, SINGLE-REPLACEMENT, DOUBLE-REPLACEMENT, OR COMBUSTION.

9. $2H_2 + O_2 \rightarrow 2H2O$

10. $2C_2H_2 + 5O_2 \rightarrow 4CO_2 + 2H_2O$
(acetylene)

11. $2HgO \rightarrow 2Hg + O_2$

12. $C_3H_5N_3O_9 \rightarrow N_2 + CO_2 + H_2O + O_2$

13. $Hg_2(ClO_3)_2 + MgBr_2 \rightarrow Hg_2Br_2 + Mg(ClO_3)_2$

14. $CuSO_4 + Zn_{(s)} \rightarrow ZnSO_4 + Cu_{(s)}$

15. Many gas grills use propane, C_3H_8, as a fuel. Write and balance the equation for the combustion of propane.

16. Flaming dishes of all kinds are made by adding liquor, warming, and lighting the food while it's in the pan. Write and balance the equation for the combustion of ethanol, C_2H_5OH.

17. Magnesium is used to make lightweight metal alloys for aircraft; however, it is flammable. Write and balance the equation for the combustion of magnesium.

18. Ammonia, NH_3, is an important fertilizer. Write and balance the combination equation for the synthesis of ammonia from its two component elements.

19. Let's say you were marooned on a planet where there was no salt, but there was a lab stocked with most of the elements. Write and balance the equation for the formation of sodium chloride. Your life depends on it!

20. Write a balanced chemical equation for the heat decomposition of plumbic oxide.

21. Write a balanced chemical equation for the the heat decomposition of silver chlorate.

22. Write a balanced chemical equation for the decomposition of hydrogen peroxide.

23. Write a balanced chemical equation for the heat decomposition of lithium nitrate.

24. Write a balanced chemical equation for the oxidation of aluminum in air.

25. Write a balanced chemical equation for the the burning of white phosphorus, P_4, in air. Assume phosphorus will lose all its valence electrons by complete oxidation.

26. Write a balanced chemical equation for the reaction of potassium and sulfur.

27. Finish and balance the following equation for a replacement reaction.

$CdCl_2 + Mg \rightarrow$

28. Finish and balance the following equation for a replacement reaction.

$Al + NiSO_4 \rightarrow$

29. Will the following displacement reaction take place, according to the Activity Series?

$Zn_{(s)} + Pb(NO_3)_{2(aq)} \rightarrow$

30. Will copper metal be oxidized by acid (H^+ ion), according to the Activity Series? (See Table 10.1)

31. Will nickel metal be oxidized by acid (H^+ ion), according to the Activity Series?

32. Use the Activity Series to write a balanced chemical equation for the reaction of lithium metal with water, if it does occur.

33. Write a balanced chemical equation for the combustion of nitrogen. Assume complete oxidation.

34. Nonmetal oxides (often called acid anhydrides) react with water. Predict the product of the reaction of CO_2 with water.

35. Dinitrogen pentoxide will also react with water to form what acid?

36. Metal oxides (often called base anhydrides) react with water to form bases containing hydroxide. Predict the product(s) of the following reactions:

a) $Na_2O + H_2O \rightarrow$

b) $BaO + H_2O \rightarrow$

37. Complete the following double-replacement reaction and underline the precipitate.

$AgNO_3 + AlF_3 \rightarrow$

38. Predict product(s) for this replacement reaction, if it occurs.

$NH_4Br + F_2 \rightarrow$

39. Complete the following double-replacement reaction and underline the precipitate.

$CaCl_2 + K_2SO_4 \rightarrow$

40. a) Write the balanced detailed ionic equation for the following reaction.

$Pb(NO_3)_2 + 2NaCl \rightarrow PbCl_2 + 2NaNO_3$

b) Write the balanced net ionic equation for the above reaction.

41. a) Finish and balance the molecular equation for the following reaction.

$BaCl_2 \quad + \quad K_3PO_4 \quad \rightarrow$

b) Write the balanced detailed ionic equation for the above reaction.

c) Write the balanced net ionic equation for the above reaction.

42. a) Finish and balance the following molecular equation.

$LiOH \quad + \quad H_2SO_4 \quad \rightarrow$

b) Write the balanced detailed ionic equation for the above reaction.

c) Write the balanced net ionic equation for the above reaction.

43. Your child eats a piece of white chalk, mostly $CaCO_3$. He then begins to burp. Calmly, you write the balanced chemical equation and determine what products are formed when chalk reacts with the hydrochloric acid in the stomach. Circle the offending product.

PRACTICE TEST II

•BALANCE THE FOLLOWING EQUATIONS:

1. $\_ScCl_3 + \_Na_3PO_4 \rightarrow \_ScPO_4 + \_NaCl$
2. $\_As + \_Cl_2 \rightarrow \_AsCl_5$
3. $\_P_4 + \_O_2 + \_Cl_2 \rightarrow \_POCl_3$
4. $\_AlBr_3 + \_K_2S \rightarrow \_Al_2S_3 + \_KBr$
5. $\_NH_3 + \_O_2 \rightarrow \_NO + \_H_2O$

•CLASSIFY THE FOLLOWING REACTIONS:

6. $ScCl_3 + Na_3PO_4 \rightarrow ScPO_4 + NaCl$
7. $As + Cl_2 \rightarrow AsCl_5$
8. $2C_2H_6 + 7O_2 \rightarrow 4CO_2 + 6H_2O$
9. $2AgNO_3 + Na_2SO_4 \rightarrow Ag_2SO_4 + 2NaNO_3$
10. $CaCO_3 \rightarrow CaO + CO_2$
11. $2H_2O_2 \rightarrow 2H_2O + O_2$

•WRITE AND BALANCE EQUATIONS FOR . . .

12. the combustion of benzene, C_6H_6 (old style lighter fluid);
13. the complete combustion of sulfur, S_8;
14. the synthesis of potassium iodide from elements;
15. the burning of sodium metal;
16. the decomposition of antimony (V) oxide, Sb_2O_5, by roasting;
17. the heat decomposition of $KClO_3$;
18. and the production of slaked lime (a base) from quicklime and water.

•FINISH AND BALANCE THE FOLLOWING EQUATIONS:

(Use the Activity Series where applicable. If no reaction occurs, just write N.R.)

19. $CuSO_{4(aq)} + Pb \rightarrow$
20. $HCl + Na \rightarrow$
21. $HgO + heat \rightarrow$
22. $Hg + HCl \rightarrow$
23. $Na_2O + H_2O \rightarrow$
24. $K + Cl_2 \rightarrow$
25. $SO_3 + H_2O \rightarrow$
26. $N_2O_3 + 2H_2O \rightarrow$
27. $F_2 + MgI_2 \rightarrow$
28. $Br_2 + KCl \rightarrow$
29. $CaCO_3 + H_2SO_4 \rightarrow$

•FOR NUMBERS 30-33, PLEASE

a) FINISH AND BALANCE THE MOLECULAR EQUATIONS,

b) WRITE BALANCED DETAILED IONIC EQUATIONS,

c) WRITE BALANCED NET IONIC EQUATIONS, AND UNDERLINE ALL PRECIPITATES.

30. a) $Pb(C_2H_3O_2)_{2(aq)}$ + $KF_{(aq)}$ →
    b)
    c)
31. a) $Ba(OH)_2$ + $Sc_2(SO_4)_3$ →
    b)
    c)
32. a) NH_4Cl + $AgNO_3$ →
    b)
    c)
33. a) $MgBr_2$ + K_2CrO_4 →
    b)
    c)

Answers to Practice Test I

1. $Pb(ClO_4)_{2(aq)} + K_2SO_{4(aq)} \rightarrow 2KClO_{4(aq)} + PbSO_{4(s)}$
2. $HC_2H_3O_{2(aq)} + NaHCO_{3(s)} \rightarrow NaC_2H_3O_{2(aq)} + H_2O_{(l)} + CO_{2(g)}$
3. $H_{2(g)} + Br_{2(g)} \rightarrow \underline{2}HBr_{(g)}$
4. $I_2 + \underline{7}F_2 \rightarrow \underline{2}IF_7$
5. $\underline{2}AgNO_{3(aq)} + BaCl_{2(aq)} \rightarrow \underline{2}AgCl_{(s)} + Ba(NO_3)_{2(aq)}$
6. $\underline{3}CaCl_{2(aq)} + \underline{2}(NH_4)_3PO_{4(aq)} \rightarrow Ca_3(PO_4)_{2(s)} + \underline{6}NH_4Cl_{(aq)}$
7. $Fe_2O_3 + \underline{3}CO \rightarrow \underline{2}Fe + \underline{3}CO_2$
8. $\underline{4}C_3H_5N_3O_9 \rightarrow \underline{6}N_2 + \underline{12}CO_2 + \underline{10}H_2O + O_2$
9. combination (Some chemists call this addition or synthesis.)
10. combustion
11. decomposition
12. decomposition
13. double-replacement (Some chemists call this metathesis.)
14. single-replacement (Some chemists call this displacement.)
15. $C_3H_8 + 5O_2 \rightarrow 3CO_2 + 4H_2O$
16. $C_2H_5OH + 3O_2 \rightarrow 2CO_2 + 3H_2O$
17. $2Mg + O_2 \rightarrow 2MgO$
18. $N_2 + 3H_2 \rightarrow 2NH_3$
19. $2Na + Cl_2 \rightarrow 2NaCl$
20. $PbO_2 \rightarrow Pb + O_2$
21. $2AgClO_3 \rightarrow 2AgCl + 3O_2$
22. $2H_2O_2 \rightarrow 2H_2O + O_2$
23. $2LiNO_3 \rightarrow 2LiNO_2 + O_2$
24. $4Al + 3O_2 \rightarrow 2Al_2O_3$
25. $P_4 + 5O_2 \rightarrow 2P_2O_5$
26. $2K + S \rightarrow K_2S$
27. $CdCl_2 + Mg \rightarrow MgCl_2 + Cd$
28. $2Al + 3NiSO_4 \rightarrow 3Ni + Al_2(SO_4)_3$
29. yes
30. no reaction
31. yes
32. $2Li + 2H_2O \rightarrow 2LiOH + H_{2(g)}$
33. $2N_2 + 5O_2 \rightarrow 2N_2O_5$
34. H_2CO_3
35. HNO_3 (Add two waters. Reduce $H_2N_2O_6$ to least whole number ratio.)

36a. NaOH (2NaOH)

36b. $Ba(OH)_2$

37. $Al(NO_3)_3 + 3\underline{AgF}$
38. $2NH_4Br + F_2 \rightarrow 2NH_4F + Br_2$

39. $\underline{CaSO_4}$ + KCl

40a. $Pb^{2+} + 2NO_3^- + 2Na^+ + 2Cl^- \rightarrow \underline{PbCl_2} + 2Na^+ + 2NO_3^-$

40b. $Pb^{2+} + 2Cl^- \rightarrow \underline{PbCl_2}$

41a. $3BaCl_2 + 2K_3PO_4 \rightarrow 6KCl + \underline{Ba_3(PO_4)_2}$

41b. $3Ba^{2+} + 6Cl^- + 6K^+ + 2PO_4^{3-} \rightarrow 6K^+ + 6Cl^- + \underline{Ba_3(PO_4)_2}$

41c. $3Ba^{2+} + 2PO_4^{3-} \rightarrow \underline{Ba_3(PO_4)_2}$

42a. $2LiOH + H_2SO_4 \rightarrow 2H_2O + Li_2SO_4$

42b. $2Li^+ + 2OH^- + 2H^+ + SO_4^{2-} \rightarrow 2H_2O + 2Li^+ + SO_4^{2-}$

42c. $2OH^- + 2H^+ \rightarrow 2H_2O$

43. $CaCO_3 + 2HCl \rightarrow CO_2 + H_2O + CaCl_2$ (with CO_2 circled) No worries, this too shall pass.

Answers to Practice Test II

1. $ScCl_3 + Na_3PO_4 \rightarrow ScPO_4 + 3NaCl$
2. $2As + 5Cl_2 \rightarrow 2AsCl_5$
3. $P_4 + 2O_2 + 6Cl_2 \rightarrow 4POCl_3$
4. $2AlBr_3 + 3K_2S \rightarrow Al_2S_3 + 6KBr$
5. $4NH_3 + 5O_2 \rightarrow 4NO + 6H_2O$
6. double-replacement (metathesis)
7. combination (synthesis)
8. combustion
9. double-replacement (metathesis)
10. decomposition
11. decomposition
12. $2C_6H_6 + 15O_2 \rightarrow 12CO_2 + 6H_2O$
13. $S_8 + 12O_2 \rightarrow 8SO_3$
14. $2K + I_2 \rightarrow 2KI$
15. $4Na + O_2 \rightarrow 2Na_2O$
16. $2Sb_2O_5 \rightarrow 4Sb + 5O_2$
17. $2KClO_3 \rightarrow 2KCl + 3O_2$
18. $CaO + H_2O \rightarrow Ca(OH)_2$
19. $CuSO_4 + Pb \rightarrow PbSO_4 + Cu$
20. $2HCl + 2Na \rightarrow 2NaCl + H_2$
21. $2HgO \rightarrow 2Hg + O_2$
22. N.R.
23. $Na_2O + H_2O \rightarrow 2NaOH$
24. $2K + Cl_2 \rightarrow 2KCl$
25. $SO_3 + H_2O \rightarrow H_2SO_4$
26. $N_2O_3 + H_2O \rightarrow 2HNO_2$
27. $F_2 + MgI_2 \rightarrow MgF_2 + I_2$
28. N.R.
29. $CaCO_3 + H_2SO_4 \rightarrow CaSO_4 + H_2O + CO_2$

30a. $Pb(C_2H_3O_2)_2 + 2KF \rightarrow \underline{PbF_2} + 2KC_2H_3O_2$

30b. $Pb^{2+} + 2C_2H_3O_2^- + 2K^+ + 2F^- \rightarrow \underline{PbF_2} + 2K^+ + 2C_2H_3O_2^-$

30c. $Pb^{2+} + 2F^- \rightarrow \underline{PbF_2}$

31a. $3Ba(OH)_2 + Sc_2(SO_4)_3 \rightarrow 3\underline{BaSO_4} + 2\underline{Sc(OH)_3}$

31b. $3Ba^{2+} + 6OH^- + 2Sc^{3+} + 3SO_4^{2-} \rightarrow 3\underline{BaSO_4} + 2\underline{Sc(OH)_3}$

31c. $3Ba^{2+} + 6OH^- + 2Sc^{3+} + 3SO_4^{2-} \rightarrow 3\underline{BaSO_4} + 2\underline{Sc(OH)_3}$

32a. $NH_4Cl + AgNO_3 \rightarrow \underline{AgCl} + NH_4NO_3$

32b. $NH_4^+ + Cl^- + Ag^+ + NO_3^- \rightarrow \underline{AgCl} + NH_4^+ + NO_3^-$

32c. $Ag^+ + Cl^- \rightarrow \underline{AgCl}$

33a. $MgBr_2 + K_2CrO_4 \rightarrow \underline{MgCrO_4} + 2KBr$

33b. $Mg^{2+} + 2Br^- + 2K^+ + CrO_4^{2-} \rightarrow \underline{MgCrO_4} + 2K^+ + 2Br^-$

33c. $Mg^{2+} + CrO_4^{2-} \rightarrow \underline{MgCrO_4}$

CHAPTER 11

Stoichiometry: Calculations Based on Chemical Equations

Chemical equations are symbolic representations of chemical processes. They present a quantitative relationship among reactants and products in chemical reactions. *Stoichiometry* is a study of quantitative chemistry based on mass.

This chapter presents stoichiometry in several different contexts. Upon first reading, all the different kinds of problems that one must solve seem confusing.

SKILLS TO ENHANCE SUCCESS IN THIS CHAPTER

The key to success in this chapter is to perceive that the seemingly different kinds of problems are variations on a single conceptual base. Realize that the mole concept applies to *any* quantitative chemistry problem. The ability to *read* a chemical equation and recognize the various mole:mole ratios presented is essential to success in solving stoichiometry problems.

CHAPTER 11 PROBLEMS

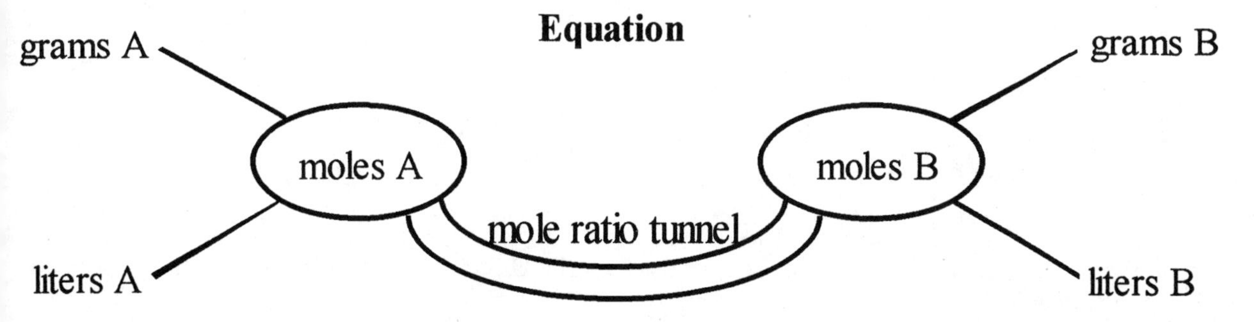

MAP OF "STOICHIOMETRIA"

We present this map to you as an alternate to the road map presented in the text.

11.1a Iron(III) oxide reacts with carbon monoxide to produce Fe_3O_4 and carbon dioxide. What are the mole ratios of this reaction?

Given: Reactants: Fe_2O_3, CO

Products: Fe_3O_4, CO_2

Need: Mole ratios of reaction

Connecting Information:

Balanced equation

Mole ratios: ratios of coefficients.

1. Write the equation line and balance the equation.

$$Fe_2O_3 + CO \longrightarrow Fe_3O_4 + CO_2$$

Multiply Fe_2O_3 by 3 and Fe_3O_4 by 2.

$$3\,Fe_2O_3 + CO \longrightarrow 2\,Fe_3O_4 + CO_2$$

The equation is balanced.

2. Write the mole ratios. (The inverse mole ratios also apply.)

$$\frac{\text{mol } Fe_2O_3}{\text{mol CO}} = \frac{3}{1} = 3 \qquad \frac{\text{mol CO}}{\text{mol } Fe_3O_4} = \frac{1}{2} = 0.5$$

$$\frac{\text{mol } Fe_2O_3}{\text{mol } Fe_3O_4} = \frac{3}{2} = 1.5 \qquad \frac{\text{mol CO}}{\text{mol } CO_2} = \frac{1}{1} = 1$$

$$\frac{\text{mol } Fe_2O_3}{\text{mol } CO_2} = \frac{3}{1} = 3 \qquad \frac{\text{mol } Fe_3O_4}{\text{mol } CO_2} = \frac{2}{1} = 2$$

11.1b. Fe_3O_4 reacts with carbon monoxide to yield FeO and carbon dioxide. Write the mole ratios of products to reactant Fe_3O_4 in this reaction.

11.2a. How many moles of Fe_3O_4 are formed from 6.00 mol Fe_2O_3 in the reaction of problem 11.1a? How many grams of Fe_3O_4 are produced?

Given: Fe_3O_4: 6.00 mol reacting with CO

Need: Mol Fe_3O_4, grams Fe_3O_4 produced

Connecting Information:

Equation for reaction: $3\,Fe_2O_3 + CO \longrightarrow 2\,Fe_3O_4 + CO_2$

Mole ratio: $\frac{2 \text{ mol } Fe_3O_4}{3 \text{ mol } Fe_2O_3}$

Molar mass Fe_3O_4 = 231.55 g/mole

1. Calculate moles Fe_3O_4 produced. (Use the mole ratio.)

$$6.00 \text{ mol } Fe_2O_3 \times \frac{2 \text{ mol } Fe_3O_4}{3 \text{ mol } Fe_2O_3} = 4.00 \text{ mol } Fe_3O_4$$

2. Calculate grams Fe_3O_4 produced.

$$4.00 \text{ mol } Fe_3O_4 \times \frac{231.55 \text{ g } Fe_3O_4}{1 \text{ mol } Fe_3O_4} = 926.2 \text{ g } Fe_3O_4 = 9.26\times10^2 \text{ g } Fe_3O_4$$

11.2b. How many moles of FeO are formed from 2.000 mol Fe_3O_4 in the reaction of problem 11.1b? How many grams of FeO are produced?

11.3a. How many moles of oxygen gas would be generated from the decomposition of 4.1 grams of mercury(II) oxide?

Given: HgO: 4.1 g

Need: Mol O_2

Connecting Information:

Equation for reaction: $2\,HgO \longrightarrow 2\,Hg + O_2$

Mole ratio: $\frac{1 \text{ mol } O_2}{2 \text{ mol HgO}}$

1. Calculate the molar mass of HgO.

$$1(200.6) + 1(16.00) = 216.6 \frac{\text{g}}{\text{mol}}$$

2. Calculate the number of moles of oxygen.

$$4.1 \text{ g HgO} \times \frac{1 \text{ mol HgO}}{216.6 \text{ g HgO}} \times \frac{1 \text{ mol } O_2}{2 \text{ mol HgO}}$$

$$= 9.5 \times 10^{-3} \text{ mol } O_2$$

11.3b. How many grams of nickel(II) sulfide would be formed from the complete reaction of 7.81 g of sodium sulfide with excess nickel(II) nitrate? The equation for the reaction is:

$$Na_2S_{(aq)} + Ni(NO_3)_{2(aq)} \longrightarrow NiS_{(aq)} + 2\ NaNO_{3(aq)}$$

11.4a. How many milliliters of 0.250 M HCl aqueous solution would be required to completely neutralize 36.5 mL of 0.100 M NaOH aqueous solution?

Given: NaOH solution: 36.5 mL, 0.100 M

Need: mL of 0.250 M HCl solution

Connecting Information:

Solution concentration:

$$\frac{0.250 \text{ mol HCl}}{1000 \text{ mL HCl}_{(aq)}} \qquad \frac{0.100 \text{ mol NaOH}}{1000 \text{ mL NaOH}_{(aq)}}$$

Equation for reaction: $HCl_{(aq)} + NaOH_{(aq)} \longrightarrow NaCl_{(aq)} + H_2O_{(aq)}$

Mole ratio, from equation: $\frac{1 \text{ mol HCl}}{1 \text{ mol NaOH}}$

Solution 1: Use of solution concentrations.

$$36.5 \text{ mL NaOH} \times \frac{0.100 \text{ mol NaOH}}{1000 \text{ mL NaOH}_{(aq)}} \times \frac{1 \text{ mol HCl}}{1 \text{ mol NaOH}} \times \frac{1000 \text{ mL HCl}_{(aq)}}{0.250 \text{ mol HCl}}$$

$$= 14.6 \text{ mL HCl solution}$$

Solution 2: Use of molarity equation.

mol solute = MV

$$V = \frac{\text{mol solute}}{M}$$

1. Calculate mol NaOH using the molarity equation.

$$(0.100 \text{ M} \times 36.5 \text{ mL})\ NaOH_{(aq)} = 3.65 \text{ mmol NaOH}$$

2. Calculate mol HCl using a mole ratio from the balanced equation.

$$3.65 \text{ mmol NaOH} \times \frac{1 \text{ mol HCl}}{1 \text{ mol NaOH}} = 3.65 \text{ mmol HCl}$$

3. Calculate volume HCl solution using the molarity equation.

$$V_{HCl} = \left[\frac{3.65 \text{ mmol}}{0.250 \text{ M}}\right]_{HCl} = 14.6 \text{ mL HCl solution}$$

We have presented the second solution in steps. You can easily combine the steps in your own presentations.

11.4b. What volume of 0.215 M sodium hydroxide aqueous solution will be needed to completely neutralize 18.4 mL of 0.169 M nitric acid? The equation for the reaction is:

$$HNO_{3(aq)} + NaOH_{(aq)} \longrightarrow H_2O_{(aq)} + NaNO_{3(aq)}$$

11.5a. Mercury and bromine react to form mercury(II) bromide. What mass of mercury(II) bromide will be formed from 67.5 g of mercury mixed with 16.0 g bromine and allowed to react at room conditions?

Given: Hg: 67.5 g

Br: 16.0 g

Need: g $HgBr_2$

Connecting Information:

Equation for reaction: $Hg_{(l)} + Br_{2(l)} \longrightarrow HgBr_{2(s)}$

Mole ratio:

$$\frac{1 \text{ mol HgBr}_2}{1 \text{ mol Hg}} \qquad \frac{1 \text{ mol HgBr}_2}{1 \text{ mol Br}_2}$$

Molar mass:

$$\frac{200.59 \text{ g Hg}}{1 \text{ mol Hg}} \qquad \frac{159.80 \text{ g Br}_2}{1 \text{ mol Br}_2} \qquad \frac{360.39 \text{ g HgBr}_2}{1 \text{ mol HgBr}_2}$$

Since the quantities of both reactants are given, this is a limiting reactant problem.

Solution 1:

1. Calculate the amount of $HgBr_2$ produced from 67.5 g Hg.

$$67.5\text{ g Hg} \times \frac{1\text{mol Hg}}{200.59\text{ g Hg}} \times \frac{1\text{ mol HgBr}_2}{1\text{ mol Hg}} \times \frac{360.39\text{ g HgBr}_2}{1\text{ mol HgBr}_2}$$
$$= 121.3\text{ g HgBr}_2$$

2. Calculate the amount of $HgBr_2$ produced from 16.0 g Br_2.

$$16.0\text{ g Br}_2 \times \frac{1\text{ mol Br}_2}{159.8\text{ g Br}_2} \times \frac{1\text{ mol HgBr}_2}{1\text{ mol Br}_2} \times \frac{360.39\text{ g HgBr}_2}{1\text{ mol HgBr}_2}$$
$$= 36.1\text{ g HgBr}_2$$

Bromine is the limiting reactant. (It yields the least amount of product.)

Amount of $HgBr_2$ formed = 36.1 g

Solution 2:

Calculate the amount of Br_2 that would react with the given amount of Hg.

$$67.5\text{ g Hg} \times \frac{1\text{mol Hg}}{200.59\text{ g Hg}} \times \frac{1\text{ mol Br}_2}{1\text{ mol Hg}} \times \frac{159.8\text{ g Br}_2}{1\text{ mol Br}_2}$$
$$= 53.77\text{ g Br}_2$$

This quantity is greater than the mass of bromine available, 16.0 g.

Bromine is the limiting reactant.

The solution is completed by the calculation based on 16.0 g Br_2, as in solution 1.

11.5b. The reaction between sulfur and chlorine is represented by the equation:

$$S_{8(s)} + 4\ Cl_{2(g)} \longrightarrow 4\ S_2Cl_{2(g)}$$

If you begin with 32.1 g sulfur and 71.0 g chlorine gas, what mass of S_2Cl_2 can you produce?

11.6a. What is the percent yield of trimethyl aluminum if 4.80 g are produced from 25.0 g of $Hg(CH_3)_2$ according to the equation given below?

$$2\ Al + 3\ Hg(CH_3)_2 \longrightarrow 2\ Al(CH_3)_3 + 3\ Hg$$

Given: $Hg(CH_3)$: 25.0 g
$Al(CH_3)_3$: 4.8 g

Need: Percent yield from reaction

Connecting Information:

Ratios:

$$\frac{230.6\text{ g Hg(CH}_3)_2}{1\text{ mol Hg(CH}_3)_2} \qquad \frac{71.98\text{ g Al(CH}_3)_3}{1\text{ mol Al(CH}_3)_3} \qquad \frac{2\text{ mole Al(CH}_3)_3}{3\text{ mol Hg(CH}_3)_2}$$

$$\text{Percent yield} = \frac{\text{actual yield}}{\text{theoretical yield}} \times 100\%$$

1. Calculate the theoretical yield of $Al(CH_3)_3$.

$$25.0\text{ g Hg(CH}_3)_2 \times \frac{1\text{ mol Hg(CH}_3)_2}{230.6\text{ g Hg(CH}_3)_2} \times \frac{2\text{ mol Al(CH}_3)_3}{3\text{ mol Hg(CH}_3)_2} \times \frac{71.98\text{ g Al(CH}_3)_3}{1\text{ mol Al(CH}_3)_3}$$
$$= 5.20\text{ g Al(CH}_3)_2$$

2. Calculate the percent yield.

$$\text{Percent yield} = \frac{4.80\text{ g Al(CH}_3)_2\text{ act.}}{5.20\text{ g Al(CH}_3)_2\text{ theo.}} \times 100\% = 92.3\%$$

11.6b. Aluminum burns in bromine. In a certain experiment, 6.0 g of Al reacts with excess bromine to yield 50.3 g $AlBr_3$. What is the percent yield in this experiment?

11.7a. P_4O_{10} reacts with water to form H_3PO_4 according to the following equation. What is the energy change in making 1.00 kg of H_3PO_4? Is the reaction exothermic or endothermic? Is heat released or absorbed during the reaction?

$$P_4O_{10(g)} + 6\ H_2O_{(l)} \longrightarrow 4\ H_3PO_{4(l)} \qquad \Delta H = -370\text{ kJ}$$

Given: H_3PO_4: 1.00 kg or 1000 g

Need: kJ

exo- or endo-, direction of energy change

Connecting Information:

Equation of reaction:

Molar mass: $98.0\ \frac{\text{g } H_3PO_4}{1 \text{ mol } H_3PO_4}$

1. ΔH is given for the reaction. This ratio applies:

$$\frac{370 \text{ kJ}}{4 \text{ mol } H_3PO_4}$$

Write the solution.

$$1000 \text{ g } H_3PO_4 \times \frac{1 \text{ mol } H_3PO_4}{98.0 \text{ g } H_3PO_4} \times \frac{370 \text{ kJ}}{4 \text{ mol } H_3PO_4}$$

$$= -944 \text{ kJ}$$

Three significant figures are warranted.

2. ΔH is negative. The reaction is exothermic.

Heat is released during the reaction.

11.7b. The energy change during combustion of 1 mole of methane is given for the reaction:

$$CH_{4(g)} + O_{2(g)} \longrightarrow CO_{2(g)} + H_2O_{(l)} \qquad \Delta H = -890 \text{ kJ}$$

Is heat released or absorbed? What mass of methane must be burned to produce 15,000 kJ?

Answers to "b" Problems

11.1b. $Fe_3O_4 + CO \longrightarrow 3\ FeO + CO_2$

$$\frac{\text{mol FeO}}{\text{mol } Fe_3O_4} = \frac{3}{1} = 3$$

$$\frac{\text{mol } CO_2}{\text{mol } Fe_3O_4} = \frac{1}{1} = 1$$

11.2b. 6.00 moles FeO; 431 g FeO

11.3b. 9.07 g NiS

11.4b. 14.5 mL NaOH solution

11.5b. 67.6 g S_2Cl_2

11.6b. 84.8%

11.7b. Heat is released. Methane required: 270 g

PRACTICE TEST I

1. How many moles of water are produced from 3.6 moles of oxygen, according to the following balanced chemical equation?

$$2H_2 + O_2 \rightarrow 2H_2O$$

2. How many moles of chlorine are needed to react with 0.25 moles of phosphorus, according to the following balanced chemical equation?

$$P_4 + 6Cl_2 \rightarrow 4PCl_3$$

3. How many moles of IF_7 are produced when 9.1 moles of fluorine reacts with excess iodine, according to the following balanced chemical equation?

$$I_2 + 7F_2 \rightarrow 2IF_7$$

4. How many moles of water are produced when 80.0 grams of hydrogen react with excess oxygen, according to the following balanced chemical equation.?

$$2H_2 + O_2 \rightarrow 2H_2O$$

5. How many grams of chlorine are needed to react with 93 grams of phosphorus, according to the following balanced chemical equation?

$$P_4 + 6Cl_2 \rightarrow 4PCl_3$$

6. How many grams of phosphorus trichloride are produced from 93.0 grams of phosphorus? (See above.)

7. How many grams of CO_2 are formed if only 40.0 g of oxygen are provided to burn the butane in the equation below?

$$2C_4H_{10} + 13O_2 \rightarrow 8CO_2 + 10H_2O$$

8. A solution of 0.800 M silver nitrate and a large amount of solid sodium chloride is available to perform the reaction below. If 25 mL of 0.800 M silver nitrate is allowed to react with excess sodium chloride, what mass of silver chloride precipitate results?

$$AgNO_3 + NaCl \rightarrow \underline{AgCl} + Na\,NO_3$$

9. Sugars are changed into ethanol by yeast fermentation, according to the equation below. What mass of ethanol, C_2H_5OH, is produced when 472 mL (about a pint) of 0.300 M dextrose (corn sugar) solution ferments?

$$C_6H_{12}O_6 \rightarrow 2C_2H_5OH + 2CO_2$$

10. What volume of oxygen is needed to burn 100. L of acetylene at the same temperature and pressure according to the equation below?

$$2C_2H_2 + 5O_2 \rightarrow 4CO_2 + 2H_2O$$

11. What total volume of gaseous products CO_2 and H_2O are produced when 100. L of acetylene burns according to the equation above?

12. What mass of calcium sulfide is produced if 25 g of each reactant are allowed to react according to the equation below?

$$Ca + S \rightarrow CaS$$

13. What mass of ammonia is produced when 100. g of nitrogen reacts with 10.0 g of hydrogen according to the equation below?

$$N_2 + 3H_2 \rightarrow 2NH_3$$

14. What weight in pounds of pure iron can be refined from one ton of ore and 400. pounds of coke (carbon)? (Use atomic weights to nearest tenths please.)

$$Fe_2O_3 + 3C \rightarrow 2Fe + 3CO$$

15. What volume of carbon dioxide is produced when 5.00 L of butane (C_4H_{10}) reacts with 32.0 L of oxygen?

$$2C_4H_{10} + 13O_2 \rightarrow 8CO_2 + 10H_2O$$

16. Given the balanced chemical equation below, 14.0 grams of aluminum and 42.0 grams of chlorine are allowed to react, and 50.4 grams of product are formed. What is the percent yield?

$$2Al + 3Cl_2 \rightarrow 2AlCl_3$$

17. How much heat will result if 65.0 grams of acetylene is burned, according to the equation below?

$$2C_2H_2 + 5O_2 \rightarrow 4CO_2 + 2H_2O + 2600\text{ kJ}$$

18. The reaction of sulfuric acid with marble is as follows. How much heat is produced by 26.0 g of $CaCO_3$?

$$H_2SO_4 + CaCO_3 \rightarrow CaSO_4 + H_2O + CO_2 + 27.0\text{ kcal}$$

19. Glucose (blood sugar) is used by the body as an energy source. Using the following equation, determine the amount of heat harvested from 270. grams of glucose.

$$C_6H_{12}O_6 + 6O_2 \rightarrow 6CO_2 + 12H_2O + 2820\text{ kJ}$$

20. For the equation below, how much heat is evolved when 5.00 grams of nitroglycerin explodes?

$$4C_3H_5O_3N_3 \rightarrow 12CO_2 + 6N_2 + O_2 + 10H_2O + 1725\text{ kcal}$$

PRACTICE TEST II

1. How many moles of hydrogen are needed to react with 2.5 moles of carbon in the following balanced chemical equation?

$$C + 2H_2 \rightarrow CH_4$$

2. Given the following balanced chemical equation, how many moles of SO_2 will react with 3.00 moles of C?

$$5C + 2SO_2 \rightarrow CS_2 + 4CO$$

3. How many moles of CO_2 will be produced when 7.0 moles of acetylene is burned with an excess of oxygen, according to the following equation?

$$2C_2H_2 + 5O_2 \rightarrow 4CO_2 + 2H_2O$$

4. How many grams of quicklime, CaO, are formed from 20. g of calcium, given the following balanced equation?

$$2Ca + O_2 \rightarrow 2CaO$$

5. Given the equation below, how many grams of hydrogen will form when 48.6 grams of magnesium react?

$$Mg + 2HCl \rightarrow MgCl_2 + H_2$$

6. For the reaction below, how many grams of CO_2 are produced if 224 grams of oxygen react?

$$2CH_3OH + 3O_2 \rightarrow 2CO_2 + 4H_2O$$

7. Given the following equation, how many grams of bromine are needed to produce 65.0 grams of product?

$$2Fe + 3Br_2 \rightarrow 2FeBr_3$$

8. How much solid gold can be precipitated out of 250. mL of a 0.500 M solution of gold (III) nitrate with an excess of aluminum, according to the following balanced chemical equation?

$$Al + Au(NO_3)_3 \rightarrow \underline{Au} + Al(NO_3)_3$$

9. What mass of plumbous choride precipitate is formed when 25 mL of .80 M plumbous acetate solution is mixed with 25 mL of 1.5 M sodium chloride solution?

$$Pb(C_2H_3O_2)_2 + 2NaCl \rightarrow PbCl_{2(s)} + 2NaC_2H_3O_2$$

10. What volume of NO_2 is produced from the decomposition of 15 L of N_2O_4 at the same temperature and pressure according to the equation below?

$$N_2O_4 \rightarrow 2NO_2$$

11. What volume of SO_3, a pollutant, is produced when 450 gallons of oxygen is provided to burn the sulfur?

$$S_8 + 12O_2 \rightarrow 8SO_3$$

12a. If 1.50 liters of gaseous iodine and 5.00 liters of gaseous fluorine react at the same temperature and pressure, what volume of product is formed? 12b. How many liters of each reactant remain?

$$I_2 + 7F_2 \rightarrow 2IF_7$$

13. How many grams of carbon dioxide are formed from 50.0 g of carbon reacting with 50.0 g of oxygen?

14. How many grams of magnesium nitride result when 63 g Mg and 41 g N_2 react?

$$3Mg + N_2 \rightarrow Mg_3N_2$$

15. What mass of water is produced when 36 grams of methanol react with 20. g of oxygen?

$$2CH_3OH + 3O_2 \rightarrow 2CO_2 + 4H_2O$$

16. According to the equation below, if a large sample of ore is treated with 95 g of carbon monoxide, and 40. grams of pure iron result, what is the percent yield?

$$Fe_2O_3 + 3CO \rightarrow 2Fe + 3CO_2$$

17. According to the equation below, if 25 kg of carbon reacts with 75 kg of sulfur dioxide, and 28 kg of carbon disulfide is formed, what is the percent yield?

$$5C + 2SO_2 \rightarrow CS_2 + 4CO$$

Answers to Practice Test I

1. 7.2 moles
2. 1.5 moles
3. 2.6 moles
4. 40.0 moles
5. 320 g
6. 413 g
7. 33.8 g
8. 2.87 g
9. 13.0 g
10. 250. L
11. 300. L
12. 45 g
13. 56.7 g
14. 1240 lb
15. 19.7 L
16. 95.8%
17. 3250 kJ
18. 7.02 kcal
19. 4230 kJ
20. 16.5 kcal (The same energy in one huge potato chip! They pack quite a punch.)

Answers to Practice Test II

1. 5.0 moles
2. 1.2 moles
3. 14 moles
4. 28 g
5. 4.00 g
6. 205 g
7. 52.7 g
8. 24.6 g
9. 5.2 g
10. 30. L
11. 300 gal

12a. 1.43 L

12b. There are 0.786 liters of iodine and 0.0 liters of fluorine remaining.

13. 68.8 g
14. 87 g
15. 15 g
16. 32%
17. 88%

CHAPTER 12

Gases

To the alchemists, the vapors that came off during experiments disappeared into the atmosphere — an "air" was being expelled. Only when these vapors were studied did chemistry become a science. Once gases such as oxygen were discovered as distinct and not just some more air, old concepts began to fall. The phlogiston theory of heat gave way to our current understanding of combustion and chemical energy.

The first understanding of the gaseous state came from Robert Brown's observation in 1827 that microscopic particles suspended in a liquid constantly move in erratic ways. Brown observed that with increased temperature, the particles move more energetically. "Brownian motion" suggested that the particles of gases were in constant motion and that temperature is a measure of the kinetic energy of the particles of a substance. After Dalton developed his atomic theory, gases became masses with widely separated and independent particles. Gases are mostly empty space. It all came together in a powerful concept called the kinetic-molecular theory (KMT).

Gases gave experimenters the notion of absolute temperature. It seems that at zero degrees absolute temperature, a gas would have no volume!

SKILLS TO ENHANCE SUCCESS IN THIS CHAPTER

The study of gases leads us very directly to the scientific method: Observation of experimental evidence led to the gas laws and kinetic molecular theory. The gas laws are not the same as "natural laws". Gas laws are actually equations resulting from systematic experimentation. Boyle, Charles, and Gay-Lussac developed the gas law equations by observing pressure, volume, and temperature behavior, making measurements, and finding a mathematical relationship between them. But the equations don't explain anything. Theories give the explanations. Theories like the kinetic molecular theory explain experimental observations on gases and the gas laws that came from these observations.

Understand the meaning of theories. They are not perfect. They represent the best model we have come up with so far. Computations based on the ideal gas law are close to reality in most cases. But no real gas follows the ideal gas equation exactly. According to calculations, at zero absolute temperature a gas would have zero volume!

Try to understand how these experimenters developed the equations. Study the figures in Sections 12.4, 12.5, and 12.6 of your text and think about everyday experiences with gases. We have mentioned proportional thinking in these notes. Understand the meaning of *inverse proportionality.*

There is nothing special or mysterious about the scientific method. Any logical person will perform it, whenever human reasoning is taking place.

CHAPTER 12 PROBLEMS

12.1a. Our atmosphere is a layer of gas about 100,000 feet thick. When you are in a commercial aircraft on a long flight, you are flying at about 35,000 feet. Would one third of air mass be below you and two thirds above you?
Answer: Much more than one third of the air mass would be below you. Air is compressible. The weight of air above compresses the air below. The air is less dense as you climb higher in the atmosphere. Air density is highest at the surface.

12.1b. If the cabin of your airplane were not pressurized, what would happen to you?

12.2a. What do we mean when we speak of an ideal gas?

Answer: An ideal gas is a gas whose behavior is described by the kinetic molecular theory.

The KM Theory makes these statements concerning the particles – atoms or molecules – of a gas:

- The particles are points of infinitesimal size.
- No force acts upon the particles from a distance.
- The particles are in constant straight-line motion.
- The particles collide in elastic collisions with one another and with the walls of their container.
- The value of average kinetic energy of the particles is the same for all gases at a given temperature and is proportional to *absolute* temperature.

According to this theory, the only forces acting upon the particles are the collisions. No "action at a distance" force such as gravity, magnetism, or electrostatic attraction affects the particles. No mutual attraction exists between the particles.

A collision is elastic when colliding masses recoil from each other with no loss of total kinetic energy. The sum of the kinetic energies of two colliding particles after collision is equal to the sum of the kinetic energies of the two particles before collision. Any gain in KE of one particle is exactly offset by loss of KE by the other particle. No amount of energy changes form. The particles do not fuse nor are they disrupted.

If one assumes any size for the particles, then one includes the statement that the distance between particles is extremely large compared to their diameters.

12.2b. Without making any measurements, how do you know that gravity does act upon the particles of a gas?

12.3a. Atmospheric pressure at sea level slightly varies around 1 atm, depending upon weather conditions. What is the cause of this pressure?

Answer: Atmospheric pressure is the weight of a column of air extending upward from the point of measurement. The base of the column is a selected area.

In the English system of measurement, the normal atmospheric pressure at sea level is 14.7 psi, pounds per square inch. The base of the column in this case is one square inch in area.

Atmospheric pressure is commonly called *barometric pressure*. The derivation of the term is obvious.

12.3b. Look at weather reports on TV. The reporter, especially if he/she is a meteorologist, will show high and low areas on a map. What are these highs and lows, and what is their cause?

12.4a. A tank of 100 L capacity contains oxygen at 150 atm, but leaks to empty while in its storage room. What volume does all of this oxygen now occupy if barometric pressure is 780 mm Hg?

Given: Oxygen: V = 100 L

P = 150 atm

Need: V at P = 780 mm Hg

Connecting Information:

Boyle's law: PV = k

1 atm = 760 mm Hg

Solution 1: Direct application of Boyle's law

1. Convert final pressure to atm.

$$780 \text{ mm Hg} \times \frac{1 \text{ atm}}{760 \text{ mm Hg}} = 1.026 \text{ atm}$$

2. Apply Boyle's law to initial, i, and final, f, conditions.

$$P_f V_f = P_i V_i$$

$$1.026 \text{ atm} \times V_f = 100 \text{ L} \times 150 \text{ atm}$$

$$V_f = \frac{100 \text{ L} \times 150 \text{ atm}}{1.026 \text{ atm}} = 1.46\times10^4 \text{ L}$$

Solution 2: Proportional reasoning

1. Convert final pressure to atm. 1.026 atm

2. Write proportions in string across line.

As pressure decreases, volume increases.
Pressure correction is greater than unity.

$$V_f = 100\ \text{L} \times \frac{150\ \text{atm}}{1.026\ \text{atm}} = 1.46\times10^4\ \text{L}$$

12.b. The tank is refilled to a pressure of 100 atm. How much oxygen is in the tank if temperature and pressure have not changed?

12.5a. What does the volume of each liter of a gas in a balloon at 100°C become if the temperature is reduced to – 40.°C without change in pressure?

Given: Gas: V = 1.00 L
T = 100.°C

Need: V at T = – 40.°C

Connecting Information:
Charles' law: V = kT

Relate final, f, volume to initial, i, volume. $V_f = V_i \frac{T_f}{T_i}$

Temperatures, absolute:
T_i = 273.15 + 100. = 373 K The least number of decimal places determines the places in the sum.
T_f = 273.15 + (– 40.) = 233 K

Convert temperatures to K before starting numerical solutions.
Be careful to properly add negative Celsius temperatures.

Solution 1: Direct application of equation derived from Charles' law

Solve for V_f from derived relationship.

$$V_f = 1\ \text{L} \times \frac{233\text{K}}{373\text{K}} = 0.625\ \text{L}$$

You can enter values first and manipulate them algebraically rather than manipulating the symbolic variables. However, be careful of transcription error if you manipulate numbers rather than variables!

Solution 2: Proportional reasoning

Write proportions in string across line.
As temperature decreases, volume decreases.
Temperature correction is less than unity.

$$V_f = 1\ \text{L} \times \frac{233\text{K}}{373\text{K}} = 0.625\ \text{L}$$

12.5b. What would be the volume at 200 °C of a liter of gas at 100°C?

12.6a. You have an aerosol can of paint in which the propellant is at a pressure of 3.0 atm at room temperature of 25°C. You inadvertently leave the can in the sun. The can's temperature rises to 95°C. What is the pressure of the propellant in the can?

Given: Gas: T_i = 273 + 25 = 298 K
P_i = 3.0 atm

Need: P_f at T_f = 273 + 95 = 368 K

Connecting Information:
Gay-Lussac law
P = kT

$$P = 3.0\ \text{atm} \times \frac{368}{298} = 3.7\ \text{atm}$$

12.6b. You see the can in the sun and hurriedly put it in your freezer at −20°C. What does the pressure in the can become?

12.7a. Why does science adopt standards such as standard temperature and pressure?
Answer: Science adopts standards for the same reason any profession or trade does:
• Enhanced work efficiency
• Communication.
If a 2", 10-20 machine screw is specified for assembly of a machine, an assembler will know what screw to use. Later, a repairman will know that he can easily obtain a replacement screw.
Do you know what the designation means? Two inches is the length of the screw from the bottom of the head. 10 is its diameter and 20 is the number of threads per inch. The screw is defined in a set of standards.
Your house is probably framed with wood. Standard 2x4s and other standard lumber sizes were used.
A set of standards is equivalent to a language common to those in the field to which the standards apply. In science, an investigator will be working at the conditions (pressure and temperature) of his/her environment. He/she will report data "reduced" to *standard conditions*.

12.7b. Why isn't a single standard for each measurement defined? Should standard pressure not be in just one of the units: atm, mm Hg or kPa?

12.8a. A family is vacationing in Smoky Mountains National Park. Children in their automobile are playing with balloons inflated with air at room temperature of 25°C and standard pressure. The vacationers drive up to Clingman's Dome at 6642 feet altitude. Atmospheric pressure there is 620 mm Hg. Temperature is freezing that day, 0°C. How much will the balloons expand or contract?

Given: Air: $T_i = 273 + 25 = 298$ K $\quad$ $T_f = 273 + 0 = 273$ K
$P_i = 760$ torr $\quad$ $P_f = 620$ torr

Need: V_f

Connecting Information:
Combined gas law

$$\left[\frac{PV}{T}\right]_i = \left[\frac{PV}{T}\right]_f$$

Solution 1: Direct application of combined gas law

$$\frac{V_f}{V_i} = \frac{P_iT_f}{P_fT_i} = \frac{760 \text{ torr} \times 273 \text{ K}}{620 \text{ torr} \times 298 \text{ K}} = 1.12$$

Expansion = (1.12 − 1.00) × 100% = 12%

Solution 2: Proportional reasoning

As pressure decreases, volume increases.
Pressure correction is greater than unity.
As temperature decreases, volume decreases.
Temperature correction is less than unity.

$$\frac{V_f}{V_i} = \frac{760 \text{ torr}}{620 \text{ torr}} \times \frac{273 \text{ K}}{298 \text{ K}} = 1.12$$

Expansion is 12%

12.8b. The vacationers inflate some new balloons for their children while at Clingman's Dome. They then drive back down the mountain to the original atmospheric conditions. How much will the balloons expand or contract? To show their children how gases behave, the vacationers inflated the new balloons to the same size as the original balloons while on the mountain. Would all balloons, both new and original, have the same volumes at the lower altitude?

12.9a. What is the volume of 0.75 mole of oxygen at 20°C and 740 torr?

Given: Oxygen: 0.75 mol
$T = 273 + 20 = 293$ K
$P = 740$ torr

Need: V at conditions

Connecting Information:
Molar volume gas at STP: 22.4 L

$$V = 22.4\ L \times 0.75 \times \frac{293\ K}{273\ K} \times \frac{760\ torr}{740\ torr} = 18.5\ L$$

This problem can be set up in many ways. Here we have emphasized the concept of molar volume of a gas. Avogadro's hypothesis slightly restated is:

Equal volumes of gases at the same temperature and pressure contain equal numbers of *moles*.

A given volume of *any* gas at constant temperature and pressure will contain a certain number of moles.

12.9b. What is the number of moles in a gas that occupies 20.2 L at 750 torr and 25°C?

12.10a. A sample of an unknown gas of volume 18.27 L has a mass of 53.95 g when measured at 22°C and 765 torr pressure. The gas is known to be an element. Identify the gas.

Given: Gas: V = 18.27 L
T = 295 K
P = 765 torr
m = 53.95 g

Need: Molar volume mass

Connecting Information:
Molar volume gas at STP: 22.4 L
Density × 22.4 L = molar mass

1. Reduce volume to that at STP.

$$V_{STP} = 18.27\ L \times \frac{273\ K}{295\ K} \times \frac{765\ torr}{760\ torr} = 17.02\ L$$

2. Calculate gas density.

$$\text{Density} = \frac{53.95\ g}{17.02\ L} = 3.170\ \frac{g}{L}$$

3. Calculate mass of 22.4 L, one mole.

$$22.4\ L \times 3.170\ \frac{g}{L} = 71.01\ g$$

4. Identify the gas of molar mass 71 g per mole.

Gas is Cl_2.

12.10b. What is the density of a gas of molar mass 36.5 g at −30°C and 1.03 atm?

12.11a. Rework problem 12.9a using the ideal gas law.

Given: Oxygen: 0.75 mole
T = 273 + 20 = 293 K
P = 740 torr = 0.9737 atm

Need: V at conditions

Connecting Information:
Ideal gas law: PV = nRT

$$R = 0.0821\ \frac{L\text{-}atm}{mol\text{-}K}$$

$$V = \frac{nRT}{P} = \frac{0.75\ mol \times 0.0821\ \frac{L\text{-}atm}{mol\text{-}K} \times 293\ K}{0.9737\ atm} = 18.5\ L$$

12.11b. Rework problem 12.9b using the ideal gas law.

12.12a. 16 g of oxygen, 28 g of nitrogen, and 8.0 g of helium are mixed in a cylinder of 80 L nominal capacity at standard temperature and pressure. What is the partial pressure of each gas?

Given: Mixture of gases at STP Standard temperature and pressure assumed.
O_2: 16 g N_2: 28 g He: 8.0 g

Need: p_{O_2}, p_{N_2}, p_{He}

Connecting Information:

Dalton's law of partial pressures
Avogadro's hypothesis

1. Calculate the number of moles of each gas and total moles.

$$O_2: \quad n = 16\text{ g} \times \frac{1\text{ mol}}{32\text{ g}} = 0.50\text{ mol}$$

$$N_2: \quad n = 28\text{ g} \times \frac{1\text{ mol}}{28\text{ g}} = 1.0\text{ mol}$$

$$He: \quad n = 8.0\text{ g} \times \frac{1\text{ mol}}{4.0\text{ g}} = 2.0\text{ mole}$$

Total moles = 0.5 mole O_2 + 1.0 mole N_2 + 2.0 mole He = 3.5 moles

2. Calculate partial pressures.

$$O_2: \quad p_{O_2} = 760\text{ torr} \times \frac{0.50\text{ mol}}{3.5\text{ mol}} = 108\text{ torr}$$

$$N_2: \quad p_{N_2} = 760\text{ torr} \times \frac{1.00\text{ mol}}{3.5\text{ mol}} = 217\text{ torr}$$

$$He: \quad p_{He} = 760\text{ torr} \times \frac{2.00\text{ mol}}{3.5\text{ mol}} = 434\text{ torr}$$

12.12b. Calculate partial pressures when the gases are mixed at 3 atm, 100°C. What effect does temperature have?

12.13a. Hydrogen has been suggested as an alternative fuel for automotive engines. Hydrogen is transported in cryogenic tanks; hardly practical for home vehicles. Hydrides have been suggested as hydrogen carriers in the fuel tank of automobiles with hydrogen-burning engines.
To investigate use of hydrides for storage of hydrogen, a chemist produces a laboratory amount of aluminum hydride by passing hydrogen through a slurry of powdered aluminum in an organic liquid. He passes 28.03 L of hydrogen gas through the slurry, which contains 53.96 g of aluminum. Reaction vessel temperature is −20°C. Pressure is 2.00 atm. How much aluminum hydride can be produced?

Given: Hydrogen: V = 28.03 L
T = 253 K
P = 2.00 atm
Aluminum: 53.96 g
Need: Theoretical yield of aluminum hydride
Connection Information:
Reaction equation: $2\,Al + 3H_2 \longrightarrow 2AlH_3$
Molar weights: Al: 26.98 H_2: 2.016 AlH_3: 30.00 g/mole

1. Calculate number of moles of aluminum.

$$Al: \quad 53.96 \times \frac{1\text{ mol}}{26.98\text{ g}} = 2.000\text{ mole}$$

2. Calculate number of moles of hydrogen.
   a. Correct volume of hydrogen to STP

$$28.03\text{ L} \times \frac{273\text{ K}}{253\text{ K}} \times \frac{2\text{ atm}}{1\text{ atm}} = \quad = 60.49\text{ L @ STP}$$

   b. Calculate number of moles.

$$n = \frac{60.49\text{ L}}{22.4\text{ L}} = 2.70\text{ mole}$$

3. Determine limiting reactant.

Hydrogen is deficient by 0.3 mole. Hydrogen is the limiting reactant.

4. Calculate theoretical yield of AlH_3

$$2.7\text{ mol }H_2 \times \frac{2\text{ mol }AlH_3}{3\text{ mol }H_2} = 1.80\text{ mole} \quad \text{that is} \quad 54.00\text{ g }AlH_3$$

12.13b. If 42.3 g of AlH_3 were recovered from this reaction, what is the percent yield?

Answers to "b" Problems

12.1b. You would die of asphyxiation quickly. Your blood cannot absorb sufficient oxygen from your lungs at the low pressures of high altitude.

12.2b. Our atmosphere is held by Earth's gravity.

12.3b. Highs and lows refer to high and low barometric pressure areas. They are caused by waves in (not on) the atmosphere and by variations of air density due to "moisture", water vapor. Many people can feel a drop in barometric pressure. They do not feel comfortable. Perhaps you can feel this drop.

12.4b. 9.75×10^3 L

12.5b. 1.27 L

12.6b. 2.5 atm

12.7b. A single standard is defined in the SI system. But more than one unit is allowed for practicality. Many different instruments and techniques are used. Different disciplines adopt different practical standards for their own work.

12.8b. The balloons contract to 89.0% of their volume on the mountain. All balloons will have the same volumes at the original conditions, within limits of measuring. Discrepancies in the calculations would be due to significant figure carrythrough.

12.9b. 0.815 mole

12.10b. $1.88 \frac{g}{L}$

12.11b. 0.815 mole

12.12b. p_{O_2} = 326 torr, p_{N_2} = 651 torr, p_{He} = 1303 torr. Temperature has no effect.

12.13b. 81% yield

PRACTICE TEST I

1. List these gases in order of their abundance in the atmosphere, starting with the most abundant:

   nitrogen   carbon dioxide   argon   oxygen.

2. The amount of water vapor in the air varies up to a maximum of about ____ .

   1%   4%   10%   25%   50-60%

3. Gases mix spontaneously with each other, a phenomenon called __________ .

4. The kinetic molecular theory states that gas samples are composed mostly of

   A) clusters or aggregates of atoms.
   B) nuclei.
   C) tightly packed molecules or atoms.
   D) empty space.
   E) high density clouds of atoms or molecules.

5. The average kinetic energy of a sample of gas

   A) increases with increasing temperature.
   B) is fixed for each given gas.
   C) is dependent on the container.
   D) is negligible.
   E) is elastic.

6. Which of these is/are approximately equal to barometric air pressure at sea level?

   1 atm   14.7 psi   760 torr   29.9 in Hg   76 cm of mercury

7. According to Boyle's Law, if the volume of a sample of gas is increased by half, what will happen to its pressure?

8. What will be the new volume if a 2.00 L balloon initially at room temperature, 25°C, is warmed to 50°C, according to Charles' law?

9. State Gay-Lussac's law.

10. An aerosol can in a cold room at 19°C contains a gas at 2.20 atmospheres of pressure. What pressure is exerted when the can is immersed in nearly boiling water at 95°C?

11. A 2.0 L helium balloon, initially at STP, is allowed to float to a height of 10 km, where the pressure is 200 torr and the temperature is – 43°C. What is the balloon's new volume?

12. A sample of carbon dioxide is at STP. The flask is then immersed in a boiling water bath. What is the new pressure within?

13. What volume is occupied by 1.5 moles of neon at STP? (Hint: Use the volume of any gas at STP.)

14. What is the density of methane at STP?

15. An unknown elemental gas is weighed and calculated to have a density of 1.783 g/L at STP. What is the gas?

16. What is the Celsius temperature of 5.0 moles of a gas at 720 torr in a 200. liter container?

17. What is the pressure of 10.0 grams of hydrogen at 25°C in a 40.0 L container?

18. Hydrogen gas (formed by the reaction of zinc and hydrochloric acid) is collected over water in a 10.0 L flask at 25°C. If 0.25 moles of hydrogen is produced, what will be the total pressure in torr within the flask of "wet" hydrogen? (The vapor pressure of water at 25°C is 23.8 torr.)

19. A familiar reaction of sodium bicarbonate and acetic acid (baking soda and vinegar) results in the production of CO_2 bubbles at a 1:1 mole ratio. A sample of 41.0 grams of sodium bicarbonate reacts with excess acetic acid, and the resulting bubbles are collected over water in a plastic bag. To what volume will the bag inflate at 25°C and 1.00 atm total pressure? Within the bag, assume a partial pressure of 24 torr for H_2O.

PRACTICE TEST II

1. About what percentage of dry air is nitrogen?

A) less than 1%
B) about 3%
C) 20%
D) 50%
E) 80%

2. Which statement about gases is false?

A) As temperature increases, volume increases.
B) Gases are mostly empty space.
C) As pressure decreases, volume increases.
D) At 0°C, the volume of a gas is theoretically zero.
E) The kinetic energy increases with increasing temperature.

3. Which is not a principle of the kinetic molecular theory?

A) Gas particles are in constant motion.
B) Gases are mostly empty space.
C) Gas particles are attracted to each other, so they collide.
D) Gas particles increase in kinetic energy when heated.
E) Energy is conserved (not lost) during a collision.

4. A 2.0-liter balloon at 1.00 atmosphere and 24°C is brought outside where it cools to 12°C at constant pressure. What is the new volume?

5. A balloon has a maximum volume of 4.0 liters. (If it gets any larger, it breaks.) Initially, the balloon is 1.5 liters in size at 25°C. Above what temperature will the balloon pop, assuming internal pressure is constant at 1.02 atm?

6. A sample of gas, initially at 10°C and 1.00 atm in a 2.0 L flask, is warmed until the pressure is 1.35 atm. What is the new Celsius temperature?

7. One mole of each of the following gases is collected at STP. Which has a greater volume than the rest? (Molecular or atomic weights given.)

A) Ar 40 g/mole
B) N_2 28 g/mole
C) NH_3 17 g/mole
D) C_2H_4 28 g/mole
E) None do.

8. What volume is occupied by 5.00 moles of oxygen (O_2) at STP?

9. What is the density of chorine gas at STP?

10. A 12.0 gram sample of helium is confined in a 5.00-liter flask at 25°C. What is its pressure?

11. Calculate the volume of 7.0 grams of nitrogen at 745 mm of mercury and 85°C.

12. A reaction of 250. mL of 3.0 M phosphoric acid with excess zinc was performed, and the hydrogen product was collected over water at 25°C and 741 torr. (The vapor pressure of water at 25°C is 24 torr.) If the volume of wet hydrogen collected was 28.0 L, what is the percent yield of hydrogen, based on the equation below?

$$2H_3PO_4 + 3Zn \rightarrow Zn_3(PO_4)_2 + 3H_2$$

13. What volume of ammonia gas could be produced from 112 liters each of hydrogen and nitrogen at STP if the reaction went to completion?

Answers to Practice Test I

1. nitrogen > oxygen > argon > carbon dioxide
2. 4%
3. diffusion
4. D
5. A
6. All of them are.
7. Pressure will decrease to half of its original value..
8. 2.17 L (You must use the Kelvin scale.)
9. At constant volume, pressure is directly proportional to temperature (in Kelvins).
10. 2.77 atm
11. 6.4 L
12. 1.37 atm
13. 33.6 L
14. 0.714 g/L
15. argon (at. wt. 39.95 g/mole)
16. 189°C (462 K)
17. 3.05 atm
18. 488 torr (464.3 torr + 23.8 torr)
19. 12.3 L

Answers to Practice Test II

1. E
2. D
3. C
4. 1.92 L
5. 795 K or 522°C
6. 109°C (382 K)
7. E (All have the same molar volume at STP.)
8. 112 L
9. 3.17 g/L for Cl_2
10. 14.7 atm
11. 7.49 L
12. 96.1% (Theoretical yield is 1.125 moles, actual yield 1.081 moles.)
13. 74.7 L

CHAPTER 13

Liquids and Solids

Liquids and solids are much more difficult to model than gases. A good explanation of the solid state had to wait for an understanding of how atoms and molecules can hold together in some tight orderly arrangement. Ionically bonded compounds are the simplest solid state substances to understand. They form hard rigid crystals having orderly lattices. Solid state theory has given us the ability to tailor-make solids with desired properties, such as the transistors in electronic devices. But our explanation didn't work for solids that do not form obvious crystals. For years scientists argued whether glass is a solid or a supercooled liquid. Glass is now believed to be a microcrystalline solid. We have a better understanding of the forces that hold particles together in soft solids and in metals, but many questions still remain.

The liquid state is still being investigated, too. Evidently, particles of liquids are held together by the same forces as those of solids, but not so tightly. Particles are continually intermingling with new neighbors. They neither have the independence of gas particles nor the interdependence of solid particles.

Most substances forming crystalline solids can exist in any of the three states of matter. You know that the substance water can exist as ice, liquid, or steam You have probably never seen any other substance in all three states of matter. Water is very unusual. At normal conditions, water undergoes abrupt obvious transitions between states, with definite energy changes. But for many substances, the distinction between the states is not as clear.

SKILLS TO ENHANCE SUCCESS IN THIS CHAPTER

This chapter presents a large number of terms and relationships that you need to sort and organize. We have to organize our information in a filing system in order to show relationships and continuity among the bits of information. Charts such as Table 13.4 of your text are extremely helpful to your understanding of the material.

Models of liquid and solid states are not as neat as the KMT model of gases. Theories, although idealizations, represent reality as best as we can perceive it at the time.

CHAPTER 13 PROBLEMS

13.1a. How many atoms of mercury are in 1.00 L of gaseous mercury at STP?

Given: 1.00 L Hg gas

Need: number of atoms in 1.00 L

Connecting Information:

Avogadro's number

Molar volume: 22.4 L

$$1.00\text{ L} \times \frac{6.02 \times 10^{23}\text{ atoms}}{22.4\text{ L}} = 2.69 \times 10^{22}\text{ atoms}$$

13.1b. What is the ratio of vapor volume to liquid volume of mercury at STP? The density of liquid mercury is 13.52 g/mL.

13.2a Review the clues to chemical bonding and the discussion of hydrogen bonds in Chapter 8. Use the information in Chapter 8 and in Sections 13.1 and 13.2 to answer the question below.

A substance is a soft solid at room temperature. It melts below the boiling point of water to form a liquid of low viscosity that does not conduct electricity. It is not soluble in water.

Is the substance a compound or an element?

What are the unit particles of the substance?
What types of *inter*particle bonds are present?
What types of *intra*particle bonds are present?
Estimate whether the molar mass is between 10 and 100 or between 100 and 1000.

Answer: The substance is a compound. Soft solid elements such as sodium react with water and conduct electricity either as solids or liquids. Solids such as charcoal are sometimes considered soft because they are easily broken but they are brittle or fractile, not soft.
The unit particles of the solid are molecules, not ions. Ionic compounds are generally soluble in water and have high melting points. They conduct electricity when melted.
Interparticle bonds are London forces.
Intraparticle bonds are covalent.
The molar mass of the compound would be *on the order of* 100, between 100 and 1000. Molecular compounds with molar masses on the order of 10 (methane, ethane, etc.) would be gases at room temperature, or liquids if they are capable of hydrogen bonding as are water and alcohols.
The compound is probably a fat or oil – an organic compound. It is obviously not ionic. Its melting point is low, so it is very likely nonpolar or only slightly polar.

13.2b. Answer the same questions for a substance that is a hard solid at room temperature. Its melting point is above 1000°C. It is soluble in water.

13.3a. What physical properties of a liquid would you measure to indicate the strength of the intramolecular bonds holding the substance in the liquid state? Explain your answer.
Answer: I would measure viscosity, surface tension, and heat of vaporization.
Heat of vaporization is a direct measure of the intermolecular bonding energy.
Viscosity, resistance to flow, and surface tension are caused by intermolecular attractions. However, a liquid can have high viscosity due to the shape and length of its molecules while having relatively low surface tension. With larger molecules, molecular weight increases and viscosity increases due to London forces. Surface tension would also increase, but not as rapidly. A lubricating oil, for example, normally has a viscosity much greater than water, but has considerably less surface tension than water.

13.3b. The SAE designation for lubricating oils used in the engine of your automobile is mentioned in your text. You probably use "multivis" (multi-viscosity) oil. Such oil is designated SAE 5W-40, for example. Why is multivis oil preferred for use in your automotive engine?

13.4a. Is the boiling of water an exothermic or endothermic reaction? Is the condensation of steam into water exothermic or endothermic? What bonds are being broken and/or formed during these processes?
Answer: Boiling, vaporization, of water requires input of energy. The reaction is endothermic.
Condensation is exothermic. Energy is released. Your hand tells you this when you scald it in steam.
Hydrogen bonds are being broken during vaporization. During condensation hydrogen bonds form. These hydrogen bonds hold water molecules together and thereby hold water in the liquid state. Steam is a gas composed of individual water molecules, a few loosely bonded to one another.

13.4b. Why does steam at 100°C give you a more severe burn than water at 100°C?

13.5a How many molecules of steam are in a liter of air above liquid water in a closed vessel at 80°C?
Given: 1 L volume of gas, 80°C.
Need: Molecules of gaseous water in equilibrium with liquid water
Connecting Information:
Table 13.1: vapor pressure water, 80°C = 355 mm Hg.
Dalton's law of partial pressures
Ideal gas law
Avogadro's number
Assume "a liter" means a unit liter with infinite sig figs, as counting numbers have.

1. Calculate the moles of water in the liter of gas.

$$\text{Ideal gas law: } PV = nRT$$

$$n = \frac{PV}{RT}$$

$$n = \frac{\left[\frac{355 \text{ mm Hg}}{760 \text{ mm Hg}} \text{atm}\right](1 \text{ L})}{\left[\frac{0.0821 \text{ L-atm}}{\text{K-mol}}\right](353 \text{ K})} = 0.01612 \text{ mol}$$

2. Calculate the number of molecules of water in the 1 L of gas.

$$\text{Number of molecules} = 1.612 \times 10^{-2} \times 6.023 \times 10^{23}$$
$$= 9.72 \times 10^{21} \text{ molecules}$$

13.5b. If the temperature of this water system were increased to 85°C, how many molecules of water would be in the vapor form?

13.6a. In calculating masses involved in chemical reactions, we speak of the molecular weight of ordinary salt, sodium chloride. Can we say that NaCl denotes a molecule of sodium chloride?
Answer: No molecule exists for NaCl in the solid state. In the gaseous state, individual NaCl units do exist, but probably most are associations of two or more NaCl units.
NaCl exists in the solid state as a crystal lattice of ions, not molecules. In a rigorous sense, an entire crystal of NaCl can be considered one molecule, since there is no distinction between the bonding between particles within the crystal. In solution, the particles are Na^+ and Cl^-.
We should use the term *formula weight* rather than molecular weight for these strongly ionic solids. In practical use for calculations it doesn't matter, since the results are the same.

13.6b. The high boiling point and high enthalpies associated with water are much greater than its molecular weight would indicate. Should we treat a molecule of water as, say, $H_{10}O_5$?

13.7a. Solid "Q" freezes at –200°C. Its specific heat is 8.375 J/g°C. How much heat is needed to raise the temperature of 20.00 g of Q from –250°C to its freezing point?
Given: 20.00 g solid, –250°C
Specific heat: 8.375 J/g°C
Need: J to −200 C

$$20.00 \text{ g} \times \frac{8.375 \text{ J}}{\text{g°C}} \times [-200°\text{C} - (-250°\text{C})] = 8.375 \times 10^3 \text{ J}$$

13.7b. How much heat is needed to melt the 20.00 g of Q if its initial temperature is –250 °C?

13.8a. We are concerned about potential "global warming". In your judgment, what would be the largest factor slowing global warming?
Answer: The oceans and polar ice caps.
The oceans are one big connected body of water, which, along with the polar ice caps, occupies 80% or so of the earth's surface. To heat all this water and melt some of the ice would require a tremendous amount of energy.
Water has the highest heats of fusion, vaporization, and specific heat of the liquid, because of strong hydrogen bonding in the liquid state. This property is of utmost importance in the evolution and maintenance of life on a planet.

13.8b. Calculate the energy required to raise 1 g of ethanol from the solid at its freezing point to the vapor at its boiling point at STP.

13.9a. Calculate the energy required to raise 1 g of water from ice at –20°C to steam at 120°C at STP.
Given: Ice at –20°C
Steam at 120°C
Need: ΔH in kJ between water states

Connecting Information:
Molar mass of water: 18.016 g
Enthalpies of water at standard conditions: Tables 13.3, 13.5, 13.6.

ΔH_{fusion}: $5.98 \frac{kJ}{mol}$ $\Delta H_{vaporization}$: $40.7 \frac{kJ}{mol}$

Specific heats: ice: $2.09 \frac{J}{g°C}$ liquid: $4.18 \frac{J}{g°C}$ steam: $1.97 \frac{J}{g°C}$

1. Calculate the energy required to heat the ice from –20°C to 0°C.

$$1 \text{ g ice} \times 2.09 \frac{J}{g°C} \times [0°C-(-20°C)] = 41.8 \text{ J} = 0.0418 \text{ kJ}$$

2. Calculate the energy required to melt 1 g H_2O.

$$1 \text{ g } H_2O \times \frac{1 \text{ mol } H_2O}{18.016 \text{ g } H_2O} \times \frac{5.98 \text{ kJ}}{1 \text{ mol } H_2O} = 0.332 \text{ kJ}$$

3. Calculate the energy required to raise liquid water from 0°C to 100°C.

$$1 \text{ g } H_2O \times 100°C \times \frac{4.18 J}{1 g \times 1°C} = 418 \text{ J} = 0.418 \text{ kJ}$$

4. Calculate the energy required to vaporize 1 g H_2O.

$$1 \text{ g } H_2O \times \frac{1 \text{ mol } H_2O}{18.016 \text{ g } H_2O} \times \frac{40.7 \text{ kJ}}{1 \text{ mol } H_2O} = 2.259 \text{ kJ}$$

5. Calculate the energy required to heat steam from 100°C to 120°C.

$$1 \text{ g steam} \times 1.97 \frac{J}{g°C} \times [120°C - 100°C] = 39.4 \text{ J} = 0.0394 \text{ kJ}$$

6. Calculate the required total energy.

$$\text{Total } \Delta H = 0.0418 \text{ kJ} + 0.332 \text{ kJ} + 0.418 \text{ kJ} + 2.259 \text{ kJ} + 0.0394 \text{ kJ} = 3.090 \text{ kJ}$$
$$= 3.09 \text{ kJ to 3 significant figures}$$

13.9b. Are the bonds holding water molecules together in the solid state (ice) stronger or weaker than the bonds holding water molecules together in the liquid state? How do you know?

Answers to "b" Problemss

13.1b. 1510:1

13.2b. The substance is an ionic compound. Interparticle forces are electrostatic ion-ion strong attractions. The molar mass of the compound could be as low as 26 for LiF or a few hundred for a compound as BaI_2.

13.3b. The viscosity, and lubricating qualities, of the oil decreases with temperature. During cold starts, the viscosity of a single SAE number oil is too large for proper circulation through the bearings and on the cylinder walls of the engine. When the engine heats, the viscosity of the lubricating oil may become too low for proper lubrication. It can get squeezed out of the bearings. Preferably, one would like the viscosity to remain constant with temperature change. The multivis oils are an approach to this. SAE 5W-40 means that the viscosity of the oil is that of an SAE 5 when cold, but that of an SAE 40 when hot. Note that a hot SAE 40 has much less viscosity than a cold SAE 40.

13.4b. The steam gives up its enthalpy of vaporization to your hand. You receive much more thermal energy from the steam than from the water.

13.5b. 1.17×10^{22} molecules

13.6b. Water molecules exist in all states; solid, liquid, and gas. Water molecules exist in solution when, for example, water dissolves in ethyl alcohol. (The water dissolves in the alcohol as well as the alcohol in the water.) The high heats associated with water are caused by strong hydrogen bonding between molecules.

13.7b. 8.81×10^3 J

13.8b. 1.42 kJ. Melting takes 109 J. Heating takes 480 J. Vaporization takes 826 J.

13.9b. Solid state bonds in ice are weaker than bonds in liquid. Evidence: $\Delta H_f \ll \Delta H_v$.

PRACTICE TEST I

1. Which of the following would you expect to be solid at room temperature?

   HCl C_6H_6 BeF_2 Br_2 H_2O PH_3

2. Which elemental metal is **not** solid at room temperature?

3. Which one of the following would you expect to be gaseous at room temperature?

   CH_4 H_2O $NaCl$ KI $C_6H_{12}O_6$ I_2

4. Which of the following would you expect to be crystalline?

   Br_2 CO $CuSO_4$ Ne N_2O CH_4

5. A sparkling solid compound is heated up to 400°C but does not melt. It does not conduct electricity unless it is dissolved in water. What type of compound is it?

6. Which exhibit primarily London dispersion forces?

   A) polar molecules
   B) long, flat nonpolar molecules
   C) hydrogen-containing molecules with N, O, or F
   D) molecules with many lone pairs
   E) British riot police

7. What intermolecular forces are primarily responsible for water's unusually high boiling point?

8. Give an example of a substance other than water that would exhibit hydrogen bonding.

9. Detergents are used to decrease what property of water?

10. Which has the highest viscosity?

    alcohol white wine coffee milk cooking oil

11. Which has the highest boiling point?

    H_2 F_2 Cl_2 Br_2 I_2

12. Which has the lowest boiling point?

    A) rubbing alcohol in Denver
    B) water in Denver
    C) rubbing alcohol in Death Valley
    D) water in Death Valley
    E) a watched pot

13. Which has the highest melting point?

    CO N_2 NH_3 H_2O Hg $NaNO_3$

14. What amount of heat in kilocalories is required to melt 10.0 pounds of ice at 0°C to water at 0°C? (The heat of fusion of water is 80 cal/g.)

15. A small candle fire vaporizes one cup of 100°C water in two hours. What amount of heat in kilocalories does the candle throw per hour? (Density of water = 1.0 g/mL; 237 mL/c ; ΔH_v = 540 cal/g.)

16. What amount of heat in kilojoules is required to convert 10.0 g of ice at -15°C to 10.0 g of steam at 120°C. (Hint: There are six steps.)

    Specific heats:

ice	2.09 J / g°C	Heat of fusion	332 J / g
water	4.18 J / g°C	Heat of vaporization	2260 J / g
steam	1.97 J / g°C		

17. Which is amorphous?

butter snowflake diamond table salt table sugar all of these

18. What type of solid contains separate discrete molecules?

19. How many potato chips at 10.0 Calories each would have to be eaten to provide enough energy to melt a roll of 50 copper pennies initially at 25°C if each weighs 3.20 grams? [ΔH_f of copper = 3.11 kcal/mole; Specific heat of copper = 0.0920 cal/g°C; Melting point of copper = 1083°C]

A) 1 small chip
B) 783 chips
C) 50 chips
D) 49,760 chips
E) 2 big chips

20. In a popular novel by Kurt Vonnegut, a new crystal structure for solid water, called "ice-nine", is discovered. It quickly covers the earth because it stays solid even when fairly warm. This ice-nine has a melting point of 40°C (104°F), so it must be heated to melt. After melting, it reverts to normal liquid water. However, below 104°F, if normal water comes in contact with a seed crystal of ice-nine, it immediately crystallizes (freezing into ice-nine). Make some guesses as to how humans could deal with the problems of a world coated with ice-nine. How could they obtain drinking water? Would the ice-nine pose a health hazard for living creatures?

PRACTICE TEST II

1. What state of matter is KNO_3?

2. Which substance exhibits dipole-dipole attraction?

CH_4 H_2O C CO_2 H_2 He

3. For which substance is hydrogen bonding most important?

NaH H_2 CH_4 HF C_6H_6

4. Which of the following is necessary if a molecule is to exhibit dipole-dipole interactions?

A) hydrogen bonding
B) permanent non-zero net dipole moment
C) zero net dipole moment
D) an elongated shape
E) N, O, or F as the central element, with hydrogens bonded to it

5. A sparkling white solid with hexagonal facets has a sharp melting point of 425°C. From this fact, one can draw the conclusion(s) that it is definitely

A) ionic.
B) covalent.
C) crystalline.
D) amorphous.
E) None of these answers aredefinitely correct.

6. Which is a gas at room temperature?

Br_2 Hg I_2 Na Al F_2

7. One of the following is a liquid at room temperature and pressure. Which is most likely in the liquid state?

A) C_5H_{12} pentane
B) C_4H_{10} butane
C) C_3H_8 propane
D) C_2H_6 ethane
E) CH_4 methane

8. Which has the highest melting point?

PCl_3 CCl_4 CF_4 BF_3 CO_2

9. What is the name of the change from gaseous to liquid state?

10. What process is used to purify liquid mixtures containing components of different volatility?

11. What name is given to the quantity of heat required to convert a mole of solid to liquid form?

A) specific heat
B) heat capacity
C) heat of reaction
D) heat of fusion
E) molar heat of solution

12. Which is a molecular solid at room temperature?

salt (NaCl) water ethanol (C_2H_5OH) silver diamond

13. What amount of heat in kilojoules is needed to melt 1.0 pound of sodium chloride, whose molar heat of fusion is 30.2 kJ?

14. Which is/are hottest?

A) simmering ethanol
B) water that just reached the boiling point
C) boiling ethanol
D) water boiling gently
E) water at a heavy boil

15. A snow cone at –4°C contains 150 g of ice, and flavoring which adds 60 Calories. Your body temperature is 37°C. Calculate the heat in kcal required to melt and warm the snow cone. Does the dessert provide enough calories to accomplish this?

Specific heats:

ice	2.09 J / g°C	Heat of fusion: 5.98 kJ / **mole**
water	4.18 J / g°C	Conversion: 4.184 J / cal

16. Give one example of a biological process that takes advantage of the relatively high heat of vaporization of water, and how it does so.

17. Why is desalination of sea water by distillation not a reasonable source of fresh water?

18. Which has the lowest boiling point?

He Ne Kr Xe Rn

19. Which has the highest melting point?

H_2O sugar ($C_{12}H_{22}O_{11}$) $CuSO_4$ propane (C_3H_8) beeswax ($C_{52}H_{104}O_2$)

20. Which of the above has the lowest boiling point?

Answers to Practice Test I

1. BeF_2
2. Hg
3. CH_4
4. $CuSO_4$
5. ionic
6. B
7. dipole-dipole interaction and hydrogen bonding
8. NH_3 or HF
9. surface tension
10. cooking oil
11. I_2
12. A
13. $NaNO_3$
14. 363 kcal
15. 64 kcal / hour
16. 30.8 kJ
17. butter
18. molecular (covalent) solid
19. E) 2 big chips (23.4 kcal = 23.4 Cal)
20. They melted the ice-nine to get drinking water, but had to keep it closed tightly to keep out seed crystals. Ice-nine would be very dangerous, because if your body came into contact with it, you would freeze solid!

Answers to Practice Test II

1. solid
2. H_2O
3. HF
4. B
5. C) crystalline (The solid is probably covalent, but it is not definitely so.)
6. F_2
7. A (B, butane, is a liquid when under pressure in a cigarette lighter, but when the valve is opened, it escapes and becomes gas under room temperature and pressure conditions.)
8. PCl_3 (It is polar, while the rest are nonpolar molecules.)
9. condensation
10. distillation
11. D
12. diamond
13. 234 kJ
14. B, D and E are tied for the hottest. The normal boiling point of water is 100°C.
15. 17.8 Cal (74.3 kJ) Yes, more than enough energy is provided by the snow cone to warm it.
16. perspiration: Living creatures can cool themselves by the evaporation of water from their skin, which removes a great deal of heat due to the high heat of vaporization of water.
17. The energy costs are too high due to the high heat of vaporization of water.
18. He
19. $CuSO_4$ (ionic)
20. propane (C_3H_8) This is a nonpolar covalent compound and is a gas at room temperature. The other choices are liquids and solids.

CHAPTER 14

Solutions

Solution is an everyday term. But you can't tell by just looking at something whether or not it is a solution. A water solution of sugar or salt looks like water. It feels like water, but you can taste the sugar or salt in it. Sometimes the substance dissolving (the *solute*) is colored and does change the appearance of the *solvent*. In some cases, the solute is tasteless. Some solutes change the texture or "feel" of the solvent water, some don't.

Many materials in nature are solutions of some sort. The air we breathe is a solution of gases. "Natural" water, particularly seawater, contains many dissolved solids and gases. Most metals used for construction and ornamentation are *alloys*, solutions of one metal in another. Even rocks may contain solutions of solids.

Life relies on control of solution concentration. Blood, serum, sap, sweat, saliva – all are solutions in water. The solutes are the important substances that are used in metabolism. The water is a carrier of nutrients, waste products, gases including oxygen and carbon dioxide, and other substances active in our metabolism.

SKILLS TO ENHANCE SUCCESS IN THIS CHAPTER

You have used the term *solution* most of your life. Again we have the situation of a common term being used in science with a precise definition. Allow your concept of solution to enlarge. Solutions are not always solids dissolved in water. The particles of a solution are molecular in size. Adding solute does affect the properties of the solvent, sometimes quite dramatically. A small amount of chromium or nickel in steel produces "stainless" steel. The solvent is not always water.

One new idea is presented in this chapter – the colloid. Particles of a colloidal suspension are much larger than molecules, but don't settle out by gravity. Living systems use colloids extensively to control internal environment.

CHAPTER 14 PROBLEMS

14.1a. Not all mixtures are solutions. How does one determine if a mixture is a solution?

Answer: A blend of small solid particles is obviously not a solution. The individual particles can be seen, using a microscope if necessary. Such mixtures are common in materials as diverse as talcum powder, soil, rocks, and concrete. The medication of medicine tablets is mixed with a material such as starch and compressed into a tablet of useful size and physical properties.

When the particles of the mixture are too small to be seen, the distinction between a solution and non-solution is not immediately obvious. You may read that a solution is a mixture that is uniform throughout. This does not merely mean uniform to the naked eye. This is the concept of a solution, but determining uniformity on a microscopic scale is not as easy as it seems. The medication and the starch in a pill are ground to very fine powders and thoroughly mixed before being formed into a tablet. The medication is considered to be uniformly distributed throughout the tablet, but it is not a solution. A solution is a mixture in which all particles, those of the solvent and those of the solute(s), are of molecular size. Since we cannot directly measure the size of the particles, we rely upon measurements of physical properties. We can make measurements as freezing point depression or boiling point elevation of the mixture to indirectly determine the size of solute particles. Visually, we expect a water solution to be completely clear. We also note the action during solvation. If a component of a mixture changes state, the mixture is a solution and the component that changes state is the solute. In dissolving in water, table salt is no longer in the solid state. It has not melted; it has dissolved. There's a big difference.

14.1b. Is fog a solution of liquid water in air? Is smoke a solution of a solid in air?

14.2a. What gas is the solvent, and what gas is the solute in air?
Answer: Applying the definition of solvent rigorously, nitrogen is the solvent and all other gases are solutes. Nitrogen has the largest concentration.
Distinction between solvent and solute comes from practice. Not just scientific practice, but pharmaceutical practice in mixing medicines, paint mixing, cooking – you name it.
Conceptually, we add the solute to the solvent, even when the solute is highly soluble and its concentration in the resultant solution is greater than the concentration of the solvent. The solute is the component that we are emphasizing, for whatever purpose, at the time. In chemistry, the solute is normally the component we wish to use as a reactant. The solvent (very often water) is a carrier of the solute.
No gas in a gas mixture is effectively either solvent or solute. Gases which do not react together will mix in all proportions. They thereby form true solutions, with no limitations on solubility.

14.2b. Isopropyl alcohol and water will mix in all proportions. Which is the solvent, and which is the solute in a mixture containing 50% of each that will be used for rubbing on the skin to reduce fever?

14.3a. Give the reason(s) that the carbonates and phosphates of ammonium, sodium, and potassium are soluble in water but the carbonates and phosphates of barium, calcium, and magnesium are not.
Refer to Table 14.2 of your text.
Answer: Many solids with doubly or triply charged ions are insoluble. Multiply charged ions attract each other more strongly than they attract water molecules.
The ammonium ion and the ions of sodium and potassium are singly charged, positively. The ions of barium, calcium, and magnesium are doubly charged, negatively. The attraction between two multiple charges is greater than the attraction between a single charge and a multiple charge.
For a substance to dissolve, the particles of the solute must separate from each other and form new bonds with particles of the solvent. The forces holding solute particles together must be less than the forces holding solute and solvent particles together.

14.3b. Sodium bicarbonate is $NaHCO_3$. Would you expect this compound to be soluble in water? Would it be more or less soluble than sodium carbonate, Na_2CO_3?

14.4a. What kinds of compounds would you expect to be soluble in liquid ammonia?
Answer: The same type of compounds that are soluble in water.
In Chapter 8 you learned of hydrogen bonding. Ammonia is highly soluble in water. Water is soluble in liquid ammonia. Ammonia can form hydrogen bonds.
Both water and ammonia are covalently bonded. The molecules of both are highly polar. Ionic compounds and compounds that are highly polar, especially compounds that will form hydrogen bonds, will dissolve in liquid ammonia.

14.4b. Medicinal alcohol is often called "rubbing alcohol". It is a mixture of isopropyl alcohol and water. You know that you will remove some of the natural oils on your skin if you cleanse it using medicinal alcohol. How does this solution, containing water, remove oils from your skin?

14.5a. A potassium nitrate solution also contains some sodium sulfate impurity. How many grams of pure potassium nitrate could be recovered from one liter of solution (before precipitation of impurity) if the solution contained 100 g KNO_3 and 20 g Na_2SO_4 per 100 g water? Refer to Figure 14.6 of your text.
Given: Solution: 1 L, 100 g KNO_3 per 100 g water, 20 g Na_2SO_4 per 100 g water
Need: Theoretical recovery of pure KNO_3
Connecting Information:
Solubility curves, Figure 14.6, text

1. Determine temperature at which Na_2SO_4 will start to precipitate.
   Reading Figure 14.6; Na_2SO_4 will precipitate at 22 °C.
2. Determine KNO_3 concentration at Na_2SO_4 precipitation.
   30 g KNO_3 per 100 g water
3. Calculate theoretical recovery of KNO_3.

$$1\text{ L} \times \frac{1000\text{ g water}}{1\text{ L water}} \times \frac{(100-30)\text{ g } KNO_3}{100\text{ g water}} = 700\text{ g } KNO_3$$

14.5b. How much KNO_3 and how much Na_2SO_4 remains in the liter of solution after the KNO_3 is precipitated?

14.6a. An aqueous solution of potassium nitrate contains 50 g of the nitrate in 50 g water. Temperature is 60°C. Some solid crystals of potassium nitrate are placed in the solution. What will happen if the temperature of this mixture is raised? Refer to Figure 14.6 of your text. Read the graph as closely as you can.
Answer: Some of the solid nitrate will dissolve.
The concentration of potassium nitrate in this solution is approximately 100 g per 100 g water. At 60°C, an aqueous solution of potassium nitrate containing 100 g of nitrate per 100 g water is saturated.

14.6b. What will happen if the temperature of the nitrate mixture is lowered?

14.7a. An isotonic solution of ordinary salt is 0.920 % salt by weight. What is the molarity of salt in this solution? Is this a dilute or concentrated solution of salt?
Given: Aqueous solution: 0.920 % salt
Need: M NaCl
Connecting Information:

$$\text{Percent by weight} = \frac{\text{g solute}}{\text{g solution}}$$

Molar mass: NaCl: 58.44 g/mole

Water density: $1.00 \frac{\text{g}}{\text{mL}}$

Assume 100 grams of solution. Assume added salt did not increase original volume solvent of water.

1. Calculate volume of 100 g solution.

$$100\text{ g solution} - 0.92\text{ g salt} = 99.08\text{ g water by weight}$$

$$\text{mass water in 100 g solution} = 99.08\text{ g}$$

$$V = M/D = 99.08\text{ g} \times \frac{\text{mL}}{1.00\text{ g}} = 99.08\text{ mL solution (same as water volume)}$$

2. Calculate moles NaCl in 1.00 L solution.

$$\frac{0.92\text{ g NaCl}}{99.08\text{ mL solution}} \times \frac{1\text{ mol NaCl}}{58.44\text{ g NaCl}} \times \frac{1000\text{ mL solution}}{1\text{ L solution}} = 0.159\text{ mol NaCl per liter solution}$$

$$M_{NaCl} = 0.159\text{ M}$$

Chemically, this solution would be considered dilute. Salt will dissolve to a maximum concentration of about 28% by weight, a fully saturated solution. But when used for medicinal purposes, an isotonic solution is neutral. A solution of greater percentage salt would be considered concentrated, since it would have to be diluted before injection into a patient. A solution of lesser percentage salt would have to be concentrated before injection. Injection of any salt solution other than an isotonic one into a patient's bloodstream is dangerous.
The terms, dilute and concentrated, are qualitative. They are relative to a standard appropriate to use. In submarines, in caissons during bridge building, in commercial diving, in any application in which workers operate in enclosed environments, oxygen concentration should be monitored. A concentration of oxygen either too high or too low is dangerous. Carbon monoxide proportions of 100 ppm (parts per million) or more in air are considered concentrated, since at this concentration a person breathing the air becomes dizzy and disoriented.

14.7b. What is the molarity of an isotonic solution of glucose, 5.5% glucose by weight?

14.8a. Years ago, one could buy either "alcohol" or "permanent" antifreeze. One was ethyl alcohol, C_2H_5OH, denatured to discourage consumption, the other ethylene glycol, $C_2H_4(OH)_2$. Compare the freezing points of solutions of 1200 g of either compound in 1 L water.
Given: Ethyl alcohol, C_2H_5OH, in water
Ethylene glycol, $C_2H_4(OH)_2$, in water
Need: Relative freezing point depressions for equal mass percentage solutions

Connecting Information:
Freezing point depression depends upon the number of "particle moles" in solution.
Molar mass: ethyl alcohol, 46.07 g ethylene glycol, 62.07 g

1. Calculate the number of moles of each compound present.

$$\text{Ethyl alcohol:} \quad 1200\text{ g} \times \frac{1\text{ mol}}{46.07\text{ g}} = 26.0\text{ mol}$$

$$\text{Ethylene glycol:} \quad 1200\text{ g} \times \frac{1\text{ mol}}{62.07\text{ g}} = 19.3\text{ mol}$$

2. Calculate the particle moles of each compound present.
Each compound produces one particle per molecule in water solution. Neither ionizes.
Number of particle moles = number of moles.
Ethyl alcohol: 26.0 particle moles
Ethylene glycol: 19.3 particle moles

3. Compare freezing point depressions.

$$\frac{\text{f.p. depression ethyl alcohol}}{\text{f.p depression ethylene glycol}} = \frac{26.0}{19.3} = 1.86$$

For the same mass per unit mass water, ethyl alcohol depresses the f.p. 1.86 times as much.

14.8b. Ethylene glycol is sweet and attractive to children and pet animals. It is poisonous. Some new antifreezes consist of propylene glycol, used in foods. Propylene glycol is $C_3H_6(OH)_2$. How much propylene glycol would be required to produce the same freezing point depression in water solution as that produced by ethylene glycol?

14.9a. Experiments to produce rain from clouds have been conducted. A cloud is "seeded" from an airplane flying above the cloud. What type of substance would be used for this seeding?
Answer: An ionic substance readily soluble in water would be used.
The water droplets forming the cloud (a fog) are held apart by surface charges on the droplets. Ions will neutralize these charges and allow the water droplets to coalesce into drops sufficiently large to fall as rain. The compound used in the experiments was sodium iodide.

14.9b. Lightning from a clear sky – the "bolt out of the blue" – is very rare. Explain why lightning is associated with dark, dense "thunderhead" clouds.

14.10a. Water normally flows out of our intestines into our bodies, carrying with it nutrients from the small, lengthy intestine. By what process do laxatives and enemas work?
Answer: Laxatives and enemas introduce a foreign substance that increases the solute concentration within the intestines. This causes water to be drawn into the intestine by osmosis; water softens the stools. A common enema consists of a solution of a phosphate in water. The high concentration of phosphate ion causes water to be drawn in. One old remedy consists of nothing more than mineral oil, used either as an enema or taken orally. It softens and lubricates.
All nutrients pass through semipermeable membranes within our bodies. The action of our kidneys, eliminating primarily waste products of protein breakdown, is that of dialysis. No special "pump" is needed. Life is not possible without osmosis.

14.10b. A physician leaves instructions to give a hospital patient 500 mL of intravenous solution isotonic in both salt and glucose. A nurse mixes 250 mL of isotonic salt and 250 mL of isotonic glucose standard solutions. Is the resultant mixture isotonic in salt and glucose? What is the approximate percent by weight of salt and of glucose in the 500 mL?

Answers to "b" Problems

14.1b. That light is scattered indicates that fog and smoke are suspensions, not solutions. Furthermore, a solid or liquid cannot be a component of a gaseous solution. Only a gas can. Water vapor is a solute in air. Odors in the air we breathe indicate that vapors of solids or liquids are dissolved in air.

14.2b. If the solution is to be used for medicinal purposes, alcohol is still considered the solute, as it is the "active ingredient".

14.3b. $NaHCO_3$ is soluble in water. The H is bonded to an O in an –OH attached to C. The presence of the –OH allows hydrogen bonding in $NaHCO_3$, causing it to be more soluble than Na_2CO_3. $NaHCO_3$ is baking soda, Na_2CO_3 is washing soda.

14.4b. Alcohol molecules contain –OH groups that can form hydrogen bonds. Medicinal alcohol is completely miscible with water. The alcohol molecule also contains a hydrocarbon group known as a methyl ($-CH_3$) group. Such hydrocarbon groups make up oils and fats. Oils are slightly soluble in alcohol. Thus, medicinal alcohol will remove small amounts of oils from a surface.

14.5b. 300 g KNO_3, 200 g Na_2SO_4.

14.6b. Solid potassium nitrate would precipitate out.

14.7b. 0.32 M glucose.

14.8b. 22.6% more by weight of propylene glycol required. Molar mass ratio; propylene glycol to ethylene glycol = 1.226.

14.9b. Lightning consists of a discharge between two oppositely charged objects. The "heavy" (dark) type of cloud contains a high concentration of water droplets. The water droplets, not the air, accumulate charge.

14.10b. The solution is not isotonic. The concentrations of salt and glucose are half the isotonic values: salt, 0.46 %; glucose, 2.8% by weight.

PRACTICE TEST I

1. Common vinegar is a 4% aqueous solution of acetic acid. Identify the solute and solvent.

2. Which pair is immiscible?

   acetic acid and water

   ethanol and water

   oil and gasoline

   gasoline and water

   copper and water

3. Fill in the <u>state</u> of matter: Brass is a _____ - _____ solution.

4. A homogeneous mixture is defined as a _________.

5. What property of water causes it to be a good solvent for ionic compounds?

6. Small alcohols (C_xH_yOH) are soluble in water because both alcohols and water are polar, but more importantly because they both exhibit what other intermolecular force?

7. Ethyl fluoride, C_2H_5F, and ethylamine, $C_2H_5NH_2$, have opposite solubilities in water although they have similar polarity and molecular weight. Which is water soluble and why?

8. Sodium acetate is very soluble in water: 100 mL of a saturated solution contains 40.5 g of this salt at 25°C. The density of the saturated solution is 1.205 g/mL. Express the solubility in terms of grams of sodium acetate per 100 <u>grams</u> of <u>water</u> instead of grams per 100 mL of <u>solution</u>. Next, calculate the molarity of the saturated solution.

9. With most solids, increasing temperature has what effect on solubility?

10. A certain mystery solution was found in the lab. A student placed a glass stirring rod in the solution, and suddenly a large amount of solid crystallized out of the solution. Was the stirring rod a magic wand, or is there a more reasonable explanation?

11. To best keep carbon dioxide gas dissolved in a carbonated beverage, what pressure and temperature combination should be maintained?

A) low pressure, high temperature
B) high pressure, low temperature
C) low pressure, low temperature
D) high pressure, high temperature
E) any pressure, low temperature

12. If 60.0 grams of glucose, $C_6H_{12}O_6$, is dissolved in water to make 500. mL of solution, what is the resulting molarity?

13. How many grams of NaOH is present in 500. mL of a 6.0 M solution?

14. What is the percent by volume of a solution of 20 mL of methanol mixed with enough water to make 80 mL of solution?

15. A solution is made by mixing 1.0 kg of cyclohexane (density 0.78 g/mL) with 1.0 kg of cyclohexanol (density 0.96 g/mL). What is the resulting percent by volume and which is the solute? (Assume volumes are additive.)

16. The maximum allowable concentration of ethyl acetate in air, according to a 1975 Federal Code, is 400 ppm by volume at 1.0 atm pressure and 25°C. Ethyl acetate, $CH_3COOC_2H_5$, is found in most nail polish removers. If 45 mL (about 3 tablespoons) of ethyl acetate spills and evaporates in a closed room of 22500 L volume (about 10 ft by 10 ft by 8 ft), will it exceed the allowable level? [Hint: Change to moles and find the volume of ethyl acetate using $V = nRT/P$.] The density of ethyl acetate is 0.900 g/mL; molecular weight is 88.0 g/mole.

17. You wish to make 10.0 L of 2.50 M acid from a stock solution of 12.5 M acid. How much stock solution will you need for this dilution?

18. Which is a suspension?

salt water   sugar water   steel   muddy water   vinegar

19. Ivory soap floats because it is soap whipped with air. What type of mixture is this?

a solution   an aerosol   an emulsion   a sol   a suspension   a colloidal dispersion

20. Certain chocolate candy bars contain oily chocolate liquor and water-based milk. In order to keep these mixed, lecithin or gums are added as _______________.

21. Sausages are often made with salt-cured meats ground and packed into a natural casing membrane made from intestinal lining. If such a sausage is soaked in water, the water will taste salty, even though the sausage remains intact. What property of the membrane allows this, and what is the name of the process?

PRACTICE TEST II

1. Which is not a solution?

A) salt water
B) HCl gas dissolved in water
C) iron filings thoroughly mixed with salt
D) solid carbon and iron melted, mixed, and cooled (steel)
E) sugar and salt dissolved in water

2. Solvation of a solute means

A) saturation.
B) trituration.
C) dynamic equilibrium.
D) dissociation and surrounding with solvent.
E) getting to heaven.

3. Which could be used as a degreasing solvent?

A) HCl, hydrochloric acid
B) CCl_4, carbon tetrachloride
C) KNO_3, saltpeter
D) CH_4, methane
E) H_2O, water

4. Which would not dissolve corn oil?

A) hexane, C_6H_{14}
B) benzene, C_6H_6
C) carbon tetrachloride, CCl_4
D) water, H_2O
E) methane, CH_4

5. What is the molarity of a solution that has 2.50 moles of KNO_3 dissolved in 5.00 L of solution?

6. How many grams of glucose ($C_6H_{12}O_6$, mol. wt. 180 g/mole) are present in 1.50 L of a 0.400 M solution?

7. What volume (in mL) of 12.00 M nitric acid is needed to prepare 100.0 mL of 0.250 M nitric acid?

8. Potassium Chloride Injection U.S.P. is a 3.00 M stock solution of KCl. You must dilute this solution to 0.100 M before administering it to a patient. To prepare 100.0 mL of the dilute solution, what volume of stock solution would you start with?

9. Which will produce the fewest ions when 1.0 mole is mixed with 1.0 L of water? (Careful!)

A) $Fe(ClO_4)_2$
B) NH_4Cl
C) $PbCl_2$
D) Na_2SO_4
E) HCl
F) $AlCl_3$

10. Which of the above will produce the most ions when 0.01 mole is mixed with 1.0 L of water?

11. Which is used as an emulsifier in dairy products like ice cream?

A) oil
B) detergent
C) gum arabic
D) CO_2
E) foam

12. Which best describes a detergent? It is

A) nonpolar.
B) moderately polar.
C) miscible with all liquids.
D) highly soluble.
E) made up of a nonpolar tail and a polar head.

13. Which is a colloid?

A) hair spray
B) shaving cream
C) jelly
D) all of these
E) none of these

14. What action could cause crystallization of a supersaturated solution?

A) shaking or blowing
B) dropping in seed crystals
C) dropping in some sand
D) trituration
E) all of these

15. "A drop in the bucket" is a common phrase. If there are 20 drops in 1.0 mL, and one of them drops into a 3.0 gallon bucket, how many ppm by volume is the one drop?

16. Which is at a dynamic equilibrium?

A) a suspension
B) a saturated solution
C) a dispersion
D) a supersaturated solution
E) all of these
F) none of these

Answers to Practice Test I

1. The solute is acetic acid; the solvent is water.
2. gasoline and water (Copper is insoluble in water, not immiscible. Immiscible referd to two liquids.)
3. solid-solid
4. solution
5. polarity
6. hydrogen bonding
7. Ethylamine, because it will hydrogen bond with water due to its hydrogen atoms attached to highly electronegative nitrogen. Ethyl fluoride will not hydrogen bond. (Its Hs are on carbon, not fluorine.)
8. 50.6 g / 100 g water   4.94 M
9. Increasing temperature increases the solubility of most solids.
10. The solution had been supersaturated.
11. B
12. 0.667 M
13. 120. g
14. 25%
15. The solution is 45% by volume cyclohexanol (solute).
16. Yes. The ethyl acetate will be at 500 ppm. (0.460 moles, 11.26 L)
17. 2.00 L
18. muddy water
19. a colloidal dispersion
20. an emulsifier or an emulsifying agent
21. The membrane is porous, or semipermeable. Salt passes through by dialysis. Water would also pass in, by osmosis.

Answers to Practice Test II

1. C
2. D
3. B
4. D
5. 0.500 M
6. 108 g
7. 2.08 mL
8. 3.33 mL
9. C (Lead (II) chloride is insoluble in water! Virtually no ions will be present.)
10. F
11. C
12. E
13. D
14. E
15. 4.4 ppm (The drop is 0.050 mL; the bucket is 11231 mL.)
16. B

CHAPTER 15

Reaction Rates and Chemical Equilibrium

Explosives react at an exceedingly rapid rate. Wood in your fireplace burns slowly as the flame propagates through the mass of the wood, but at the point of burning, the reaction is proceeding quickly. We see wood rotting and cars rusting slowly. All are reactions, but with different *rates*.

The rate of a reaction depends greatly upon the *conditions* of the reaction. If hydrogen and oxygen are mixed and ignited, they react explosively. In a fuel cell, the hydrogen and oxygen react slowly, producing electricity. The conditions of the reaction are being controlled.

Reversible reactions reach a state in which they appear to have stopped. However, the reaction is still operating. On the molecular level, two opposing reactions are taking place at the same rate. We call these reactions the forward and reverse reactions. The reacting mass has reached a state of *equilibrium*.

All reactions need a "kick" to get started. The kick is energy input to the reacting mass. It's called *activation energy*. For the spontaneous reactions going on about us at all times, the kick is small and immediately available from thermal energy at normal temperatures. Other reactions require larger kicks. A *catalyst* lowers the intensity of the kick required to get a reaction started.

SKILLS TO ENHANCE SUCCESS IN THIS CHAPTER

Potential reactants, whether atoms or molecules, must actually collide in order to react. They cannot react at a distance. They must be able to move. To react, the orbitals of two atoms must overlap. Electrons in these orbitals must switch around between the atoms being bonded. Review Chapter 5 on electron structure and Chapter 8 on bonding.

A condition that favors (increases the likelihood of) particle collision increases the rate of a reaction. Higher pressure squeezes the particles closer together. Higher temperature allows them to move around more quickly. An excess of one reactant puts more particles of that reactant in the reacting mass. Increasing the surface area of a solid or liquid reactant means more of its particles are exposed to collision with particles of another reactant.

A reaction produces products. If one of the products leaves the reacting mass as a gas or a precipitate, the reaction continues in the *forward* direction until completion. If all products stay within the reacting mass, they can become reactants themselves, for the *reverse* reaction. If this is the case, the reaction is said to be *reversible*, and a two-way arrow is used. As the reaction proceeds, the concentration of the reactants decreases and the rate of the forward reaction decreases. During this same time, the concentration of the products increases and the rate of the reverse reaction increases. When the two rates are equal, the reaction apparently ceases. The reaction is at *equilibrium*.

CHAPTER 15 PROBLEMS

15.1a. Criticize this statement: Reactions occurring in nature have no activation energy. After all, they are *natural*.
Answer: Reactions occurring in nature have high activation energy. If they did not, they would proceed uncontrollably. Sugar would spontaneously convert to water and carbon dioxide. Proteins would reform into other proteins willy-nilly. There would be no order or differentiation of tissues. But the reactions can't be *too* energy costly, or the organisms might not survive, especially during infancy or lean times. Enzymes have evolved as a way to manipulate these high-activation reactions, making them easier to start up, yet they still remain under biological control. The storage of high-energy molecules is also extremely important to a cell. To be safe, the energy molecule cannot be too reactive. It, too, must be controlled. Without high activation energy, the balanced cycles of Earth would randomize. Life could not exist.

15.1b. What is the source(s) of the activation energy for reactions occurring on Earth?

15.2a. The effect of pressure is not included as a factor controlling reaction rate in Section 15.2. How does pressure affect reaction rate? Would an increase of pressure increase the reaction rate of solid reactants?
Answer: Pressure is a measure of concentration of a gas. Pressure applied at constant temperature to a gas decreases the volume of the gas. Thus the concentration of gas increases and the reaction rate increases.
An increase in pressure would not appreciably affect the reaction rate of solids, since solids are for the most part incompressible. The concentration of the solids would not change. [The effect of increasing pressure on solid reactants would only be noticeable at very high pressures. Nothing is totally incompressible.]
Think in terms of particles per unit volume, concentration, in determining reaction rates.

15.2b. Gas mixture A, at 700 mm Hg pressure, is 20% oxygen by volume. Gas mixture B, at 500 mm Hg pressure, is 28% oxygen by volume. In which gas mixture is oxygen the more concentrated?

15.3a. Certain freons, used as propellants in spray cans of many types and as the operating fluid in refrigeration and air-conditioning units, have been banned. Your new car's air conditioner has a "safe" propellant. Why were the freons banned?
Answer: Because freons destroy ozone. Ozone "holes" (areas where ozone is lacking) have been found in the stratosphere. Ozone high in the atmosphere absorbs UV radiation from the sun. This radiation, if it penetrated to the surface of the earth, would be highly damaging to life. Ozone protects by absorbing the ultraviolet light, *hv*, in this reaction, shown as a mechanism of two steps:

$$hv + O_2 \longleftrightarrow O + O$$
$$O + O_2 \longleftrightarrow O_3$$

The concentration of ozone in the stratosphere is a few ppm by mass or particle count, but as low as this concentration is, it does shield us from the UV.
Blame for the ozone holes has been put on these freons. In the 70s, one million tons of these substances were produced annually. The freons are chlorofluorocarbons. They are highly stable and nonpoisonous compounds — the reason that they attained such extensive use. They ascend into the stratosphere and destroy the ozone by these reactions:

$$hv + CFCl_3 \longleftrightarrow CFCl_2 + Cl\bullet$$
$$Cl\bullet + O_3 \longleftrightarrow ClO + O_2$$
$$O + ClO \longleftrightarrow Cl\bullet + O_2$$

The atomic chlorine is a *free radical*. A free radical is an atom or group of atoms with an unpaired electron. The dot after the atomic symbol represents the unpaired electron. The chlorine radical is a reactant in the second reaction and is a product in the third reaction. It is not used up and can continue to recycle through the overall reaction.

15.3b. The reaction of hydrogen and chlorine, although highly exothermic and extremely rapid, does not occur if hydrogen and chlorine are mixed in the dark even at room temperatures. Just a little ordinary light will ignite the mixture explosively. What does this imply concerning the activation energy of the reaction of hydrogen and chlorine?

15.4a. For this reaction at equilibrium in a closed vessel,

$$4\,HCl_{(g)} + O_{2(g)} \longleftrightarrow 2\,H_2O_{(g)} + 2\,Cl_{2(g)} + 27\text{ kcal}$$

how would equilibrium be affected by each of these individual changes?

- increased temperature
- increased pressure
- increased volume of vessel
- addition of water.

Answers:
- Increased temperature: The equilibrium shifts toward the reactants, to the left, away from the heat term.
The reaction is exothermic. The energy released can be treated as a product.
We can write the equation this way:

$$4\,HCl_{(g)} + O_{2(g)} \longleftrightarrow 2\,H_2O_{(g)} + 2\,Cl_{2(g)} + 27\text{ kcal}$$

Adding product shifts the equilibrium toward the reactant side of the equation.

• Increased pressure: The equilibrium shifts toward the products, right.
All reactants and products are gases in this equation. Five moles of gas react to produce four moles of gas. Write in the letter P on the left, the side with more gas molecules. That side has more pressure.

$$\mathbf{P} + 4\,HCl_{(g)} + O_{2(g)} \longleftrightarrow 2\,H_2O_{(g)} + 2\,Cl_{2(g)} + 27\text{ kcal}$$

Equilibrium will shift away from the P. Increasing pressure shifts equilibrium toward the side of fewer moles of gas, to the right.

• Increased volume: The equilibrium shifts left, toward the reactants, to replace the lost pressure.
Increasing the volume decreases the concentration of the gases.
Increased volume decreases pressure, so the equilibrium will shift toward the letter P.

• Addition of water: The equilibrium shifts left, toward the reactants, away from water.
Water gas is one of the products. Adding more water increases the concentration of water.

Try this approach. Whatever we do to a system at equilibrium, it will shift to "undo" what we did. If we add more reactant, the equilibrium will shift away from the reactant side (toward the right), making more product to balance out the added reactant. If we add more product, it will shift away from that product (back to the left). Removing a participant means the equilibrium will shift toward the lost species. If the reaction produces energy, treat energy like a product. "Place" pressure on the side having the most gas molecules. Adding heat or pressure will cause a shift away from the heat or pressure term. Reducing either heat or pressure will cause a shift toward the lost item. This is essentially a detailed restatement of LeChâtelier's principle: that the reaction responds to relieve stress.

15.4b. For this reaction carried out in a closed vessel,

$$5\,CO_{(g)} + I_2O_{5(s)} \longleftrightarrow I_{2(g)} + 5\,CO_{2(g)}$$

how would equilibrium be affected by these changes?

- increased pressure
- addition of catalyst
- addition of I_2O_5
- decrease in volume of vessel.

15.5a. Enzymes are the catalysts of life. How does an enzyme work?
Answer: Enzymes are extremely specific, each affecting reaction of a single compound, called the substrate. A substrate molecule fits into an active site on the enzyme. The model is that of a *lock and key*. For particles to react, they have to collide with sufficient energy and in the right orientation. An enzyme ensures the right orientation by fitting the substrate molecule, the key, into its lock. It also supplies energy by electric charge, moving the electron pairs in the substrate molecule, and by changing shape to strain bonds within the substrate molecule.
The complicated organic reactions of metabolism are extremely slow and produce a mixture of closely related products when performed in a laboratory. Enzymes increase the forward rate of these reactions by hundreds of thousands of times and result in only the desired product.
Enzymes also turn off reactions to control the production of substances within the body according to demand. Our metabolism is entirely controlled by enzymes.

15.5b. The "catalytic converter" in your automobile contains a *surface catalyst.* Describe the action of this type of catalyst by analogy with the action of an enzyme. Note that no enzymatic lock-and-key mechanism exists for these catalysts. Consider the elements used for these catalysts.

15.6a. What is the concentration at equilibrium of each gas in a 10 L container containing 1 mole of gaseous water? The K_{eq} for water decomposition is 6.0×10^{-30} at about 450°C.
Given: Water decomposition $K_{eq} = 6.0 \times 10^{-30}$
Need: $[H_2O]$, $[H_2]$, $[O_2]$
Connecting Information:
Reactant and products are gases at temperature of experiment.

1. Write equation of decomposition.

$$2\,H_2O \longleftrightarrow 2\,H_2 + O_2$$

2. Write K_{eq} equation.

$$K_{eq} = \frac{[H_2]^2[O_2]}{[H_2O]^2}\frac{mol}{L}$$

3. Calculate equilibrium concentrations in terms of molar decomposition.

Let x = moles water decomposed.

$$[H_2]: \quad \frac{1\,x\text{ mol}}{10\text{ L}} = 0.1x\text{ M}$$

$$[O_2]: \quad \frac{0.5\,x\text{ mol}}{10\text{ L}} = 0.05x\text{ M}$$

$$[H_2O]: \quad \frac{(1-x)\text{ mol}}{10\text{ L}} = (0.1-x)\text{ M}$$

4. Calculate the water dissociated at equilibrium.

$$K_{eq} = 6.0 \times 10^{-30} = \frac{(0.1x)^2(0.05x)}{(0.1\text{-}x)^2}\text{M}$$

Simplify the algebraic manipulation by noting that the decomposition is very small.
Assume that $(0.1 - x) = 0.1$ within precision limits of calculation.

$$K_{eq} = 6.0 \times 10^{-30} = \frac{(0.1x)^2(0.05x)}{(0.1)^2}\text{M}$$

$$6.0 \times 10^{-30} = 0.05\,x^3\text{ M}$$

$$120 \times 10^{-30} = x^3$$

$$4.932 \times 10^{-10}\text{ M} = x$$

5. Calculate the concentrations at equilibrium.

$$[H_2] = 0.1x = 4.932 \times 10^{-11}\text{ M}$$

$$[O_2] = 0.05x = 2.466 \times 10^{-11}\text{ M}$$

$$[H_2O] = 0.1\text{ M}$$

6. Check simplification by calculating K_{eq} from the concentrations.

$$\frac{(4.932{\times}10^{-11})^2(2.466{\times}10^{-11})}{0.1^2} = 5998 \times 10^{-33} = 6 \times 10^{-30}$$

Four significant figures were carried for the purpose of checking the answer. This number of significant figures is not warranted in the final reported answer.

7. Restate the concentrations at equilibrium considering significant figures.

$[H_2]$: 4.9×10^{-11} M

$[O_2]$: 2.5×10^{-11} M

$[H_2O]$: 0.1 M

15.6b. 1.00 mole of ICl is placed in a 1.00 L container and allowed to dissociate. The equation of dissociation is:

$$2\,ICl \longleftrightarrow I_2 + Cl_2$$

The equilibrium constant $K_{eq} = 2.2 \times 10^{-3}$ for this reaction at the temperature of the experiment. Write the equilibrium expression and an algebraic equation for this reaction in terms of x, where x is the concentration of I_2 at equilibrium. Estimate the relative size of x. Can you use the same simplification as in Problem 15.6a, Step 4?

Answers to "b" Problems

15.1b. The sun is the ultimate energy source for reactions occurring on the surface of the earth. You know that we, as animals, live in reciprocity with the plants. The plants produce our food through photosynthesis and return oxygen to us. Weather derives its energy from the sun. Petroleum products are decayed plant material. In the earth's crust, the source of energy can also be residual heat from the earth's formation or radioactivity.

15.2b. The concentration of the oxygen is equal in both gases, 140 mm Hg.

15.3b. The activation energy is high. It is higher than available from thermal energy, but not so high as to require intense ultraviolet. The hydrogen and chlorine can be mixed under a red light such as a photographic darkroom safelight.

15.4b. Increased pressure: equilibrium shifts left toward reactants. Pressure of products is greater than pressure of reactants. P is on the right side. Increasing pressure will shift equilibrium to the left.
Addition of catalyst: no effect upon equilibrium.
Addition of I_2O_5: no effect upon equilibrium. I_2O_5 is a solid. Addition of more of it does not change its concentration in the reaction mixture.
Decreased volume: equilibrium shifts left toward reactants. Decreasing volume increases pressure.

15.5b. Reactant particles are *adsorbed* on the surface of the catalyst. They are thereby brought in close contact with one another. The reactant gases effectively condense on the surface of the catalyst. These catalysts are transition elements from group VIIIB which have closely spaced outer orbitals and, especially at high temperatures, many loosely bound electrons, enhancing the formation of radicals and intermediates in the reaction mechanisms by which the pollutants are oxidized.

15.6b. $K_{eq} = \frac{[I_2][Cl_2]}{[ICl]^2} = \frac{x^2}{(1-2x)^2}$ How large is x? Try dropping the 2x to estimate the value of x.
$K_{eq} = 2.2 \times 10^{-3} = x^2$ So, $x = 0.047$
x = about 5% of initial ICl concentration, therefore too large to be dropped as in the previous problem.
No, you cannot make the same simplification.

PRACTICE TEST I

1. A reaction occurs only if

A) a catalyst is present.
B) a collision occurs.
C) the reactants are warm.
D) effective collisions occur.
E) the reaction is exothermic.

2. The activation energy is

A) the minimum amount of energy necessary for successful collisions.
B) the energy needed to properly orient the molecules.
C) the difference in energy between reactants and products.
D) the net energy liberated during the reaction.
E) the net energy absorbed during the reaction.

3. Why does a catalyst speed up a reaction?

4. You are a process manager at a chemical plant making vitamins. What ways could be explored to speed up production?

5. In an exothermic reaction,

A) the net energy of the products is higher than the net energy of the reactants.
B) the net energy of the products is lower than the net energy of the reactants.
C) the activation energy is high.
D) the activation energy is low.
E) a catalyst is not effective in increasing the rate of reaction.

6. Why does increasing the temperature so profoundly affect the rate of reaction? (Select any that apply.)

A) It increases the frequency of collisions.
B) It causes products to stabilize.
C) It increases the energy of participants so more collisions will reach the activation energy.
D) The mechanism for this phenomenon is unknown.
E) It lowers the activation energy.

7. In the reaction of zinc metal with aqueous hydrochloric acid, hydrogen gas is the fizzing product. With which reactants will the most vigorous fizzing be observed at a given temperature?

A) 1.0 g mossy zinc chunks and 100 mL of 6.0 M HCl

B) 1.0 g zinc dust and 50 mL of 6.0 M HCl
C) 1.0 g zinc granules and 100 mL of 6.0 M HCl

8. An ordinary reversible reaction is held at constant temperature for several hours. What will happen? (Select all that apply.)

A) It will reach equilibrium.
B) The reaction will have stopped by then.
C) The forward reaction runs at a higher rate than the reverse reaction.
D) All collisions will cease.
E) Collisions will still occur, but none will be effective.

9. The fact that a system at equilibrium, if altered, will compensate for the alteration and return to a new equilibrium is called ____________________.

10 – 14. For the next five questions, use the equation below:

(LEFT) $P + 2B_{(s)} + 3F_{2(g)} \leftrightarrow 2BF_{3(g)} + 45\ kJ$ (RIGHT)

Predict how equilibrium will be affected if . . .

10. temperature is increased at constant pressure.
11. volume is decreased at constant temperature.
12. boron is added.
13. BF_3 is added.
14. a catalyst is added.

15. Write the balanced chemical equation represented by the following equilibrium expression.

$$\frac{[OF_2]^2}{[O_2]\ [F_2]^2}$$

PRACTICE TEST II

1. A successful collision requires which of the following? (Select any that apply.)
    A) That the right molecules (or atoms) collide.
    B) That the reaction vessel be of sufficient size.
    C) That the molecules (or atoms) have sufficient energy when they collide.
    D) That the molecules (or atoms) be properly oriented when they collide.
    E) That a catalyst be present.

2. For the diagram below, what does the dotted line represent?

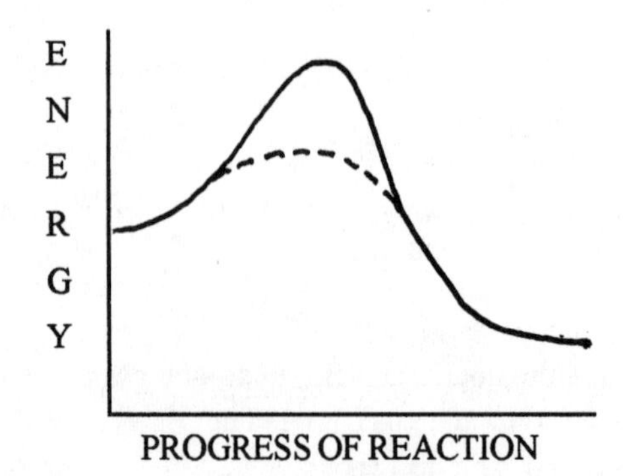

Questions 3 – 6.

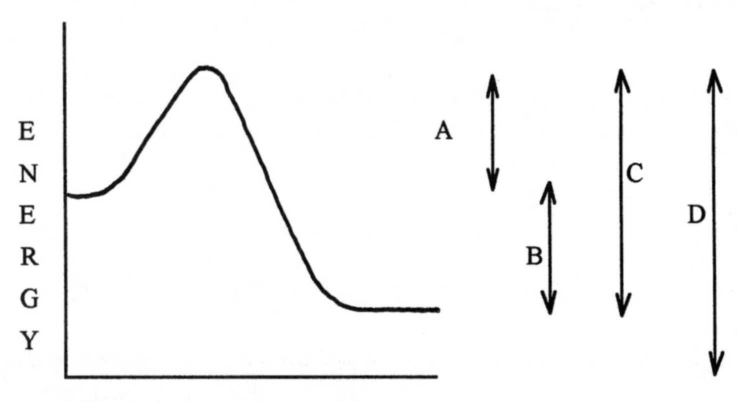

3. Which arrow represents the net energy of reaction?
4. Which arrow represents the activation energy of the forward reaction?
5. Which arrow represents the activation energy of the reverse reaction?
6. Is the reaction endothermic, exothermic, or thermoneutral?

7. A catalyst
   A) increases the rates of both forward and reverse reactions.
   B) increases the rate of the forward reaction only.
   C) causes the equilibrium to shift to the right (favoring products).
   D) supplies energy to reactants.
   E) inhibits the reverse reaction.

8. Increasing the temperature of a reaction by 10°C
   A) doubles the rate, in general.
   B) increases the number of collisions.
   C) increases the probability of a successful collision.
   D) increases the rate of the reverse reaction.
   E) all of the above

9. Which of the following is/are true of equilibrium? (Select all that apply.)
   A) No reactions are occurring.
   B) The reactants are all used up.
   C) The rate of the forward reaction equals the rate of the reverse reaction.
   D) The concentrations of reactants and products remain constant.
   E) The number of collisions is at a maximum.

For the next six questions (10 – 15.), consider the following gaseous equilibrium:

(LEFT) $N_2 + 3H_2 \rightarrow 2NH_3 + 45\text{ kJ}$ (RIGHT)

Use LeChâtelier's Principle to predict the shift in equilibrium upon . . .

10. addition of H_2 at constant pressure.
11. addition of NH_3 at constant pressure.
12. an increase in pressure at constant temperature.
13. a decrease in temperature at constant pressure.
14. addition of a catalyst.
15. an increase in volume at constant temperature.

16. Write the equilibrium expression for the following gaseous reaction:

$$2N_2 + 5O_2 \leftrightarrow 2N_2O_5\ .$$

Answers to Practice Test I

1. D
2. A
3. A catalyst lowers the activation energy of the reaction, so the collisions are more often successful.
4. You could find a catalyst, increase the temperature of the reaction, increase concentrations of reactants, and/or increase the surface area by grinding or powdering solid reactants.
5. B
6. A and C
7. B (The HCl concentrations are all the same, so only the zinc particle size matters.)
8. A
9. LeChâtelier's Principle
10. left
11. right
12. no effect (A solid does not show up in the equilibrium expression.)
13. left
14. no effect (not applicable)
15. $O_2 + 2F_2 \leftrightarrow 2OF_2$

Answers to Practice Test II

1. A, C, and D
2. The dotted line shows the reaction pathway with a catalyst present.
3. B
4. A
5. C
6. exothermic
7. A
8. E
9. C and D
10. right
11. left
12. right
13. right
14. no effect
15. left
16. $\frac{[N_2O_5]^2}{[N_2]^2[O_2]^5}$

CHAPTER 16

Acids and Bases

You have used the word "acid", and probably have heard words like "basic" or "alkaline", "pH", "neutral", and so forth. This chapter will give you a much better understanding of these terms.

Pure water is *neutral*. But pure water is not found in nature. Most liquid water is either acidic or basic. The water that you drink is either slightly acidic or basic. Wines are slightly acidic, even when sweet. Some foods are highly acidic, especially the sour foods like lemons, vinegar, sour cream, spinach, rhubarb, etc. Have you ever tried to eat cranberries without sweetener? If you are a gardener, you know that you have to control the acidity of the soil in which you grow ornamentals or food crops. Each plant requires the soil in which it grows to be of a certain acidity. We are concerned with "acid rain". Rain is always acidic! The water of rain has dissolved carbon dioxide and other substances that produce acids when dissolved in water. In fact, most reactions that we see occurring on the surface of the earth involve acids and bases. Life itself relies upon a delicate acid-base balance. Our stomachs produce acid strong enough to be damaging to mucous membranes. Hyperacidity can cause ulcers in the stomach and esophagus. Our gallbladder produces bile, a base to neutralize the stomach acid. The internal pH of our blood and body fluids is tightly controlled. If this control is lost, metabolic reactions shut down or produce the wrong products. Acid-base chemistry is extremely important and is used in many different fields of study.

SKILLS TO ENHANCE SUCCESS IN THIS CHAPTER

You will be expanding your concepts of acids and bases in this chapter. As before, you will be reading words that are part of everyday language but in chemistry are more precisely defined.

Acids and bases are conceptually opposites. When we wish to describe a solution, whether acidic or basic, we simply measure its acidity, since a certain acidity (pH) level implies (and is mathematically linked to) a certain level of base. Most reactions involve merely a net transfer of the H^+ ion from the acid to the base. Keeping this simplified view in mind will help you when the equations look intimidating. It's all a matter of a little proton, H^+.

The chemist's concept has expanded to include different definitions of acid and base. These different theories are not in conflict, but are just a slightly different way of looking at acid and base behavior. You will see, again, that theory – idealization – helps us to understand the real world.

CHAPTER 16 PROBLEMS

16.1a. Identify the Arrhenius acid in each step of the following stepwise ionization of phosphoric acid.

1) $H_3PO_4 + HOH \longleftrightarrow H_3O^+ + H_2PO_4^-$
2) $H_2PO_4^- + HOH \longleftrightarrow H_3O^+ + HPO_4^{2-}$
3) $HPO_4^{2-} + HOH \longleftrightarrow H_3O^+ + PO_4^{3-}$

Answer: In step 1), H_3PO_4 is the Arrhenius acid. In step 2), $H_2PO_4^-$. In step 3), HPO_4^{2-}.

16.1b. Write the steps for the ionization of sulfuric acid. Identify the Arrhenius acid in each step.

16.2a. We call hydrochloric acid a strong acid. Is a 0.01 M solution of hydrochloric acid "weak"?

Answer: At first glance, the question seems silly. We have discussed the confusing nature of qualitative terminology before. Let's try to be as "chemical" as we can now.

HCl, more properly hydrogen chloride, is a gas. HCl dissolved in water is called hydrochloric acid. It is a strong acid because it is highly ionized in water.

The 0.01 M solution of HCl is weak by any standards. It is weaker than the HCl solution produced in our stomachs. It is the solution that is weak, not the HCl itself, which is an acid. We should probably be more careful with our terminology. We should call the solution dilute, not weak.

We could even argue whether we should call HCl an acid. HCl itself does not turn litmus red. But HCl dissolved in water forms the hydrogen ion [H^+], as Arrhenius said, or the hydronium ion [H_3O^+], as we say today. It is the ion that turns litmus red. Then which really is the acid, HCl or [H_3O^+]? Your text describes an Arrhenius acid as a compound that releases hydrogen ions in water. On this basis, HCl is an acid. So also is the hydronium ion.

16.2b. Boric acid is used for medicinal purposes. The formula for boric acid is $B(OH)_3$. The formula looks like the formula of a base! Show that boric acid is indeed an acid.

The structure for boric acid is:

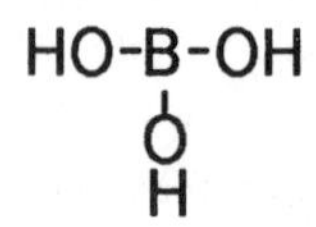

16.3a. Is potassium hydroxide an Arrhenius acid or a base, and is it strong or weak? Answer the same questions for potassium oxide.

Answer: KOH gives OH^- in water – it is an Arrhenius base. Since it ionizes completely, it is strong. K_2O forms KOH in water, which in turn releases hydroxide ions. It is a strong Arrhenius base.

16.3b. Is ammonium hydroxide an Arrhenius acid or a base, and is it strong or weak?

16.4a. Show, using chemical equations, what produces the taste of a carbonated soft drink when first opened, and explain what happens when allowed to stand.

Answer: A carbonated drink is bottled in an atmosphere of carbon dioxide at greater than 1 atm pressure. The carbon dioxide dissolves in the water of the drink, forming "soda" water. The reaction is:

$$CO_2 + H_2O \longleftrightarrow H_2CO_3$$

The carbonic acid formed ionizes:

$$H_2CO_3 \longleftrightarrow H^+ + HCO_3^-$$

which can also be written:

$$H_2CO_3 + H_2O \longleftrightarrow H_3O^+ + HCO_3^-$$

The hydronium ion causes a sour taste.

Little carbonic acid, H_2CO_3, forms, and not all of that ionizes. H_2CO_3 cannot exist as a free compound. It exists only in water solution. For H_2CO_3 to remain in solution, CO_2 gas must be in solution. The equilibrium of the decomposition reaction is far to the right:

$$H_2CO_3 \longleftrightarrow H_2O + CO_2\uparrow$$

As CO_2 gas escapes when the drink is left open to the atmosphere, the concentration of CO_2 in solution becomes so low that eventually all but a small percentage of H_2CO_3 has decomposed.

The acidity of the drink continually decreases during this time. The sour taste eventually disappears.

16.4b. Write the equation for the reaction of baking soda with vinegar. What is the gaseous product?

16.5a. You have seen TV commercials that advertise the effectiveness of an antacid. One antacid label lists these quantities of bases per dose: 200 mg aluminum hydroxide and 200 mg of magnesium hydroxide. Many other ingredients, including flavorings, are of course present.

How much hydrochloric acid of 0.100 M concentration will one dose of this antacid neutralize?

Given: $Al(OH)_3$: 200 mg
$Mg(OH)_2$: 200 mg

Need: Stoichiometric HCl, 0.100 M

Connecting Information:
Molar mass: $Al(OH)_3$: 78.00 g $Mg(OH)_2$: 58.33 g

1. Write equations for neutralization of bases.

$$Al(OH)_3 + 3HCl \longrightarrow AlCl_3 + 3HOH$$
$$Mg(OH)_2 + 2HCl \longrightarrow MgCl_2 + 2HOH$$

2. Calculate HCl required for each base and add.
Use millimoles, since reactants are in mg.

$$200\text{ mg Al(OH)}_3 \times \frac{1\text{ mmol Al(OH)}_3}{78.00\text{ mg Al(OH)}_3} \times \frac{3\text{ mmol HCl}}{1\text{mmol HCl}} \times \frac{1\text{mLHCl}}{0.100\text{mmolHCl}}$$

= 76.92 mL HCl to neutralize $Al(OH)_3$

$$200\text{ mg Mg(OH)}_2 \times \frac{1\text{ mmol Mg(OH)}_2}{58.33\text{ mg Mg(OH)}_2} \times \frac{2\text{ mmol HCl}}{1\text{mmol HCl}} \times \frac{1\text{ mL HCl}}{0.100\text{ mmol HCl}}$$

= 68.58 mL HCl to neutralize $Mg(OH)_2$

3. Add the quantities of HCl.

76.92 mL + 68.58 mL = 145.5 mL HCl

16.5b. Another antacid contains 400 g of $Mg(OH)_2$ per dose. How much of the 0.100 M HCl would one dose of this antacid neutralize?

16.6a. Can we produce one acid from another? Complete the following equation and show how the Brønsted-Lowry theory indicates to us that we can. Label the conjugate pairs.

$$HNO_3 + F^- \longrightarrow$$

Answer:

HNO_3	+	F^-	$\longrightarrow$	NO_3^-	+	HF
$Acid_{\#1}$		$CB_{\#2}$		$CB_{\#1}$		$Acid_{\#2}$

Hydrofluoric acid can be formed from nitric acid by this reaction. A weaker acid like HF can be formed from a stronger acid like HNO_3 by reacting the stronger acid with the conjugate base of the weaker acid (fluoride). The fluoride ion comes from a compound such as NaF in solution.

16.6b. What is the result of mixing sodium carbonate and aqueous hydrochloric acid? Write the equation of the reaction. Identify the conjugate pairs.

16.7a. The structure of our bodies, including bones, is built from protein. Protein is composed of chains of amino acids. Show using the Lewis diagram that an amino acid is a Lewis base.
A modified Lewis structural formula for an amino acid can be written:

```
         R
   ..    |    =O
 H N — C — C
   H          \OH
```

Not all electrons are indicated. R represents a hydrocarbon (hydrogen and carbon) assembly of some form.
Answer: The nitrogen has a lone pair of electrons. They can be donated and allow the amino acid to function as a Lewis base.

16.7b. You also recognize the –COOH at the other end. This is the (organic) acid end of an amino acid. Amino acids react with one another and link up head to tail. What would you predict is the *type* of reaction that accomplishes this? How do amino acids join in chains to form proteins?

16.8a. The concentration of acetic acid in vinegar can range from 1% to 5%, by weight. Vinegar of 3% by weight acetic acid has a hydrogen ion concentration of 3.0×10^{-3} molar. What is the hydroxide ion concentration?
Given: $[H^+] = 3.0 \times 10^{-3}$ M
Need: $[OH^-]$
Connecting Information:
Ion product constant of water: $K_w = 1.0 \times 10^{-14}$

Basis of calculation: constancy of K_w

$$[H^+] \times [OH^-] = 1.0 \times 10^{-14}$$

$$3.0 \times 10^{-3} \times [OH^-] = 1.0 \times 10^{-14}$$

$$[OH^-] = \frac{1.0 \times 10^{-14}}{3.0 \times 10^{-3}}$$

$$= 3.3 \times 10^{-12}\text{ M}$$

16.8b. If the vinegar is 1.5% by weight acetic acid, the hydrogen ion concentration is 2.1×10^{-3} molar. What is the hydroxide ion concentration in this weaker vinegar?

16.9a. What is the pH of the 3% vinegar of problem 16.8a?
Given: $[H^+] = 3.0 \times 10^{-3}$ M
Need: pH
Connecting Information:
$pH = -\log [H^+]$

$$\begin{aligned} pH &= -\log(3.0 \times 10^{-3}) \\ &= -[\log 3.0 + \log 10^{-3}] \\ &= -[0.477 + (-3)] \\ &= -0.477 + 3 \\ &= 2.5 \end{aligned}$$

16.9b. What is the pH of the 1.5% vinegar of problem 16.8b?

16.10a. What is the acidity of a solution of ammonium chloride in water? Prove your conclusion using Table 16.3 in your text.
Answer: By Table 16.3, we would judge the solution to be acidic.
NH_4Cl is a salt of NH_4^+ (cation of a weak base) and Cl^- (anion of a strong acid).
Considering these hydrolysis reactions:

1) $NH_4^+ + H_2O \longleftrightarrow NH_3 + H_3O^+$
2) $Cl^- + H_2O \not\longrightarrow HCl + OH^-$

Reaction 1 will proceed because NH_3 is a weak base. Hydrolysis of water will produce acid.
Reaction 2 will not proceed to the right, therefore it will not change the pH of water by hydrolysis.
Considering this reaction:
3) $NH_4^+ + Cl^- \not\longrightarrow HCl + NH_3$
Reaction 3 will not proceed to the right because it is forming a stronger acid (HCl over NH_4^+) and a stronger base (NH_3 over Cl^-). This will not occur spontaneously.
The relative strengths of the acids, HCl and NH_4^+, and the corresponding relative strength of the conjugate bases, Cl^- and NH_4^+, is the determining factor.

16.10b. What is the acidity of a solution of $NaC_2H_3O_3$ in water? Prove your conclusion.

16.11a. What salt would be used to make an acetic acid buffer? Write the reactions that exert pH control upon addition of acid H^+ and base OH^- to this buffer.
Answer: The salt used in an acidic buffer introduces into the solution the conjugate base of the acid of the buffer. In an acetic acid buffer, the salt would be sodium acetate. The control reactions in a $HC_2H_3O_2/C_2H_3O_2^-$ buffer are:
•neutralization of added acid by:

$$C_2H_3O_2^- + H^+ \longleftrightarrow HC_2H_3O_2$$

•neutralization of added base by:

$$HC_2H_3O_2 + OH^- \longleftrightarrow C_2H_3O_2^- + H_2O$$

16.11b. Why is a mixture of HCl and NaCl not a buffer?

16.12a. The equivalence point of an acid-base titration is the point at which the acid and base exist in stoichiometric ratio to one another. What is the difference between the end point and the equivalence point? Is the pH always equal to 7 at the equivalence point? Give an example for a titration whose end point would not have a neutral pH.

Answer:
End point
The end point of a titration occurs when the indicator used changes color, signaling a change in pH. If this indicator is a chemical one, the indicator requires only a small amount of acid or base to effect a change in its color.

Equivalence point

Take for example the neutralization reaction between hydrochlo ric acid and sodium hydroxide:

$$HCl + NaOH \longrightarrow NaCl + HOH$$

The equivalence point exists when the ratio of NaOH to HCl is one-to-one:

$$\frac{[NaOH]}{[HCl]} = 1$$

Since this is a reaction between a strong acid and a strong base, the pH at equivalence is 7, neutral. There is no hydrolysis of the products; they are neutral. But this is not always the case.

If the titration is that of a weak base with a strong acid, or a strong base with a weak acid, the pH is not 7 due to hydrolysis of one of the product ions formed. An example would be the titration of NaOH and acetic acid or any other weak acid. To determine pH at equivalence, check the products for hydrolysis.

16.12b. Estimate the pH at the equivalence point of a titration of vinegar with sodium hydroxide. Is the resulting solution acidic, basic, or neutral? Explain your reasoning. What indicator would you pick from Table 16.6 for this titration? What color change would you watch for?

Answers to "b" Problems

16.1b. 1) $H_2SO_4 + HOH \longrightarrow H_3O^+ + HSO_4^-$ The Arrhenius acid is H_2SO_4.

2) $HSO_4^- + HOH \longleftrightarrow H_3O^+ + SO_4^{2-}$ The Arrhenius acid is HSO_4^-.

16.2b. Boron has only three valence level electrons. It can form stable compounds without the necessity of an octet of electrons. But the B atom in a compound such as $B(OH)_3$ boric acid can accept an electron pair to form an octet. An ion of $B(OH)_4^-$ is formed in water solution:

```
[     H     ] −
      O
   HO-B-OH
      O
[     H     ]
```

The equation would be $B(OH)_3 + HOH \longleftrightarrow B(OH)_4^- + H^+$. Because hydrogen ion is produced, $B(OH)_3$ is an Arrhenius acid.

16.3b. A solution of ammonium hydroxide releases OH^- ions into the water. It is an Arrhenius base. But it is a weak base, since most of the molecules do not ionize. Ammonium hydroxide is another way of writing aqueous ammonia, NH_3. Ammonia remains mostly as NH_3 in water. It is a weak base.

16.4b. $HC_2H_3O_2 + NaHCO_3 \longleftrightarrow NaC_2H_3O_2 + H_2CO_3$

The H_2CO_3 goes on to produce water and carbon dioxide gas, which causes the bubbling.

16.5b. 137.2 mL HCl

16.6b. $Na_2CO_3 + 2HCl \longrightarrow H_2CO_3 + 2NaCl$ pairs: $Na_2CO_3//H_2CO_3$ HCl//NaCl

16.7b. The amino acids have a base end and an acid end. They link up end to end through a neutralization type reaction.

16.8b. 4.8×10^{-12} M

16.9b. 2.7

16.10b. Acetate is the anion of a weak acid, acetic acid. It will hydrolyze, producing a basic solution. Sodium is the cation of the strong base sodium hydroxide. It will not hydrolyze; it is neutral. A solution of sodium acetate will be basic.

16.11b. Neither the cation nor the anion can hydrolyze. There is no equilibrium. Only weak conjugate pairs can form buffers.

16.12b. The pH at the equivalence point is basic, approximately 9. The acetate ion hydrolyzes to a basic solution. Pick thymol blue as an indicator. Watch for color change from yellow to blue.

PRACTICE TEST I

1. Acids taste _______ and turn litmus _________.
2. Which donates an electron pair?
   A) an Arrhenius acid
   B) an Arrhenius base
   C) a Lewis acid
   D) a Lewis base
   E) a Brønsted-Lowry acid
3. Which is a weak base?
   A) HF B) NaCl C) NH_3 D) NaOH E) NO_3
4. Which is a strong base?
   A) HOH B) NH_3 C) $Mg(OH)_2$ D) KOH E) CH_3OH
5. Give the name and formula of a strong diprotic acid.
6. Sour milk contains
   A) ammonia.
   B) sodium hydroxide.
   C) $Mg(OH)_2$.
   D) sodium chloride.
   E) lactic acid.
7. Based on taste alone, which of the following contains the most acid?
   A) limes B) milk C) baking chocolate D) water E) coffee
8. Write the balanced chemical equation for the dissociation of potassium hydroxide in water.
9. Write the balanced chemical equation for the hydrolysis of the weak base ethylamine, $CH_3CH_2NH_2$, in water. [Hint: Use the hydrolysis of ammonia as a model.]

PREDICT THE PRODUCTS IN THE FOLLOWING AQUEOUS REACTIONS.

(You need not balance the equations unless you require practice.)

10. $HF + Na \rightarrow$
11. $Zn + HNO_3 \rightarrow$
12. $Al + H_2SO_4 \rightarrow$
13. $HCl + Ba(OH)_2 \rightarrow$
14. $H_3PO_4 + 3KOH \rightarrow$
15. $HC_2H_3O_2 + K_2O \rightarrow$
16. $2HCl + Na_2CO_3 \rightarrow$
17. $CaCO_3 + H_2SO_4$ (acid rain) $\rightarrow$
18. $FeCl_3 + NaOH \rightarrow$
19. $F^- + H_2O \rightarrow$
20. What is the conjugate base of HCl?
21. What is the conjugate base of H_2O?
22. What is the conjugate base of H_2CO_3?
23. What is the conjugate acid of NH_3?
24. What is the conjugate acid of $C_2H_3O_2^-$?
25. What is the conjugate acid of HS^-?
26. Which is a Lewis acid?
    A) BF_3 B) NH_3 C) H_2O D) H_2 E) NH_4^+
27. An aqueous solution has a hydroxide ion concentration of 1.0×10^{-10} M. What is its hydrogen ion concentration?

28. An aqueous solution has a hydroxide ion concentration of 2.5×10^{-3} M. What is its hydrogen ion concentration?

29. A solution that has a pH of 8.0 would have what hydrogen ion molarity?

30. What is the pH of a solution whose hydrogen ion concentration is 2.0×10^{-5} M?

31. What is the pH of a solution whose hydroxide ion concentration, $[OH^-]$, is 5.0×10^{-12} M?

32. See #31 above. Is the solution acidic, basic, or neutral?

33. What is the pH of a solution whose pOH is 6?

34. See #33 above. Is the solution acidic, basic, or neutral?

35. What is the hydrogen ion concentration of blood plasma, whose pH is 8.2?

36. Is an aqueous solution of $NaHCO_3$ acidic, basic, or neutral?

37. Is an aqueous solution of KF acidic, basic, or neutral?

38. What is the name for a solution which resists change in pH?

39. What should be added to NaF to make a solution that tends to hold pH constant when acids or bases are added?

40. Dilute vinegar (pH = 4) and ammonia (pH = 10) can both be used to wash windows. But vinegar is

A) better on salads.
B) a million times more acidic than ammonia.
C) sixty times more acidic than ammonia.
D) six times more acidic than ammonia.
E) about twice as acidic as ammonia.

41. How many mL of 0.25 M NaOH will it take to neutralize 20.0 mL of 0.15 M acetic acid?

42. A 10.0 mL vinegar sample takes 40.0 mL of 0.20 M KOH to titrate to end point. What is the molarity of acetic acid in the vinegar?

PRACTICE TEST II

1. You have a solution of $HC_2H_3O_2$. It would it taste ________, feel ________, and turn litmus ________.

A) sour slimy blue
C) sour slippery red
B) sour wet red
D) bitter wet red
E) mellow slippery yellow

2. You have a solution of pH = 10. It would it taste ________, feel ________, and turn litmus ________.

A) wet sour red
C) wet bitter blue
B) slippery bitter blue
D) bumpy dry red
E) slimy sweet green

3. Which is a strong acid?

A) chloric acid
C) acetic acid
B) hydrosulfuric acid
D) nitric acid
E) Schwarzenegger acid

4. You need a base to neutralize excess stomach acid. Which could safely be used, NaOH or $Mg(OH)_2$, and why?

5. Which contains base that helps give it its characteristic taste?

A) coffee B) sourdough bread C) lemonade D) vinegar E) sugar

6. What is the formula of lye, used to unclog drains?

7. In the United States, this chemical (used in car batteries) is produced in larger quantities than any other.

8. Write the balanced chemical equation for the hydrolysis of ammonia in water.

9. Ammonia is classified as what?

PREDICT THE PRODUCT(S) IN THE FOLLOWING AQUEOUS REACTIONS.

(You need not balance the equations. Some coefficients have been provided to help you.)

10. $HNO_3 + K \rightarrow$

11. $H_2SO_4 + Ca \rightarrow$

12. $H_3PO_4 + Mg \rightarrow$

13. $HI + LiOH \rightarrow$

14. $H_2S + 2KOH \rightarrow$

15. $6HNO_3 + Fe_2O_3 \rightarrow$

16. $HC_2H_3O_2$ (vinegar) + $NaHCO_3$ (baking soda) $\rightarrow$

17. $HCl + ZnS \rightarrow$

18. $KOH + Co(NO_3)_2 \rightarrow$

19. $C_2H_3O_2^- + H_2O \rightarrow$

20. What is the conjugate base of HNO_3?

21. What is the conjugate base of H_2CO_3?

22. What is the conjugate base of SO_4^{2-}?

23. What is the conjugate acid of HPO_4^{2-}?

24. What is the conjugate acid of H_2O ?

25. What is the conjugate acid of $C_6H_5NH_2$?

26. Which is a Lewis base? [Hint: Draw Lewis electron-dot structures]

A) H_2 B) BH_3 C) NH_3 D) NH_4^+ E) H^+

27. A solution has a hydrogen ion concentration of 1.0×10^{-7} M. What is its hydroxide ion concentration?

28. See #27 above. Is the solution acidic, basic, or neutral?

29. An aqueous solution has a hydrogen ion concentration of 5.0×10^{-11} M. What is its $[OH^-]$ concentration? Is it acidic, basic, or neutral?

30. A solution has a pH of 6. What is its hydrogen ion concentration?

A) 6×10^{-10} M B) 6×10^{-7} M C) 1×10^{-8} M D) 1×10^{-6} M E) pretty darn high

31. See #30 above. Is the solution acidic, basic, or neutral?

32. A solution has a hydrogen ion concentration of 4×10^{-6} M. What's the pH? Is it acidic, basic, or neutral?

33. What is the pOH of a solution whose hydrogen ion concentration is 1.0×10^{-3} M? Is it acidic, basic, or neutral?

34. An aqueous solution has a hydroxide ion concentration of 2.5×10^{-4} M. What is its pH?

35. What is the hydrogen ion concentration of stomach acid, whose pH is 2.3 ?

36. Is an aqueous solution of Na_2CO_3 acidic, basic, or neutral?

37. Is an aqueous solution of NH_4Cl acidic, basic, or neutral?

38. What salt should be used along with acetic acid to form a buffer solution?

39. An important carbon-containing buffer system in the blood consists of what two species?

40. A solution of strychnine, a weak base, has a pH around 10. "Dr Pepper" soda pop has a pH around 5. Strychnine (rat poison) is ...

A) ten times more basic than Dr Pepper.
B) twice as basic as Dr Pepper.
C) fifty times more basic than Dr Pepper.
D) 100,000 times more basic than Dr Pepper.
E) recommended by Doctors, twice as often.

41. What is the molarity of an unknown monoprotic acid if a 10.00 mL sample of unknown takes 12.50 mL of 0.2500 M NaOH to titrate it?

42. It takes 45.0 mL of 0.20 M NaOH to titrate 90.0 mL of HCl to end point. What is the molarity of the acid?

Answers to Practice Test I

1. sour ... red
2. D
3. C
4. D
5. sulfuric acid, H_2SO_4
6. E
7. A
8. $KOH \rightarrow K^+ + OH^-$
9. $CH_3CH_2NH_2 + H_2O \leftrightarrow CH_3CH_2NH_3^+ + OH^-$
10. $\rightarrow NaF + H_2$
11. $\rightarrow Zn(NO_3)_2 + H_2$
12. $\rightarrow Al_2(SO_4)_3 + H_2$
13. $\rightarrow H_2O + BaCl_2$
14. $\rightarrow 3H_2O + K_3PO_4$
15. $\rightarrow H_2 + 2KC_2H_3O_2$
16. $\rightarrow NaCl + CO_2 + H_2O$
17. $\rightarrow CaSO_4 + CO_2 + H_2O$
18. $\rightarrow Fe(OH)_3 + 3NaCl$
19. $\rightarrow HF + OH^-$
20. Cl^-
21. OH^-
22. HCO_3^-
23. NH_4^+
24. $HC_2H_3O_2$
25. H_2S
26. A (None of the others have "room" to accept another electron pair.)
27. 1.0×10^{-4} M
28. 4.0×10^{-12} M
29. 1.0×10^{-8} M
30. 4.7
31. 2.7
32. acidic
33. 8
34. basic
35. 6.3×10^{-9} M
36. basic (due to hydrolysis of HCO_3^-)
37. basic (due to hydrolysis of F^-)
38. a buffer
39. HF
40. B
41. 12 mL
42. 0.80 M

Answers to Practice Test II

1. B
2. B
3. D
4. $Mg(OH)_2$, because NaOH is a strong base and too caustic. (It would burn.)
5. A
6. NaOH
7. What is sulfuric acid? (Ding!)
8. $NH_3 + H_2O \leftrightarrow NH_4^+ + OH^-$
9. a Brønsted-Lowry base
10. $\rightarrow KNO_3 + H_2$
11. $\rightarrow CaSO_4 + H_2$
12. $\rightarrow Mg_3(PO_4)_2 + H_2$
13. $\rightarrow H_2O + LiI$
14. $\rightarrow 2H_2O + K_2S$ (coefficients optional)
15. $\rightarrow 3H_2O + 2Fe(NO_3)_3$
16. $\rightarrow NaC_2H_3O_2 + CO_2 + H_2O$
17. $\rightarrow ZnCl_2 + H_2S$
18. $\rightarrow Co(OH)_2 + KNO_3$
19. $\rightarrow HC_2H_3O_2 + OH^-$
20. NO_3^-
21. HCO_3^-
22. HSO_4^-
23. $H_2PO_4^-$
24. H_3O^+
25. $C_6H_5NH_3^+$
26. C (Ammonia has an electron pair that can be donated.)
27. 1.0×10^{-7} M
28. neutral
29. 2.0×10^{-4} M basic
30. D
31. acidic
32. 5.4 acidic
33. pOH is 11 acidic (pH is 3)
34. 10.4 (Did you get 1.04? See the 01 exponent in your calculator display? That's 1.04×10^1.)
35. 5.0×10^{-3} M
36. basic (due to hydrolysis of CO_3^{2-})
37. acidic (NH_4^+ is a proton donor: $NH_4^+ \leftrightarrow H^+ + NH_3$)
38. $NaC_2H_3O_2$ (sodium acetate)
39. H_2CO_3 and HCO_3^-
40. D
41. 0.3125 M
42. 0.10 M

CHAPTER 17

Oxidation and Reduction

Perhaps you heat your home with gas from pipelines. The gas is burned in your furnace, producing heat. Natural gas coming through pipelines around the nation is methane. The reaction of burning is:

$$CH_4 + 2\,O_2 \longrightarrow CO_2 + 2\,H_2O$$

The oxygen comes from the air. We say the methane has been *oxidized*, because it reacted with oxygen. Methane is a *reduced* substance, meaning it is ready to be oxidized. All fuels are reduced substances. (The terminology does not apply to nuclear "fuels".)

Think of the oxygen–carbon dioxide cycle within which we live synergistically with plants. We take in sugar by eating plants, and oxygen by inhaling air. We produce and eliminate carbon dioxide and water. Plants take in carbon dioxide and water and produce sugar in their cells. They eliminate oxygen. We are oxidizing the sugar to carbon dioxide. Plants are reducing the carbon dioxide to sugar.

The concept of oxidation and an opposite reduction is a powerful tool in understanding many chemical reactions, not just those of fuels or those in which oxygen is a reactant.

SKILLS TO ENHANCE SUCCESS IN THIS CHAPTER

Oxidation-reduction chemistry is often referred to as **redox** chemistry, for good reason. When there is a loss of electrons (oxidation), there must also be a gain (reduction). Electrons cannot be created or destroyed. They are only transferred. Oxidation and reduction always occur at the same time, hence the term redox. The term reduction seems a bit backwards, because it is a gain in the number of electrons. But remember, electrons are negative particles. Gaining electrons means that the charge will become more negative.

Think of the number line. Reduction and oxidation are opposites.

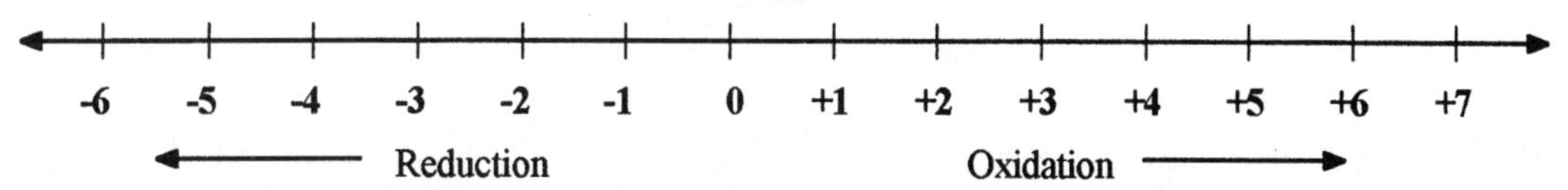

Movement to the left on the number line is reduction; a gain in electrons means going more negative.

You should review *electronegativity* in Section 8.3, and the Activity Series in Section 10.8. Review *oxidation numbers* and the rules for assigning them in Section 6.7. *Oxidation number* is the ionic charge of an atom or the *apparent charge* assigned to an atom. The atoms within covalently bonded compounds are not ions, but do have a formal charge. Oxidation number at first seems to be nothing more than a bookkeeping system, but it is more than that. It helps us understand where electron pairs of bonds are going. That's what chemical reactivity is all about.

The concept of oxidation helps us understand the majority of chemical reactions. We can separate a reaction into two *half-reactions*; one oxidation, the other reduction. One half-reaction cannot occur without the other, but the two half-reactions can be separated in space. They do in our voltaic and electrolytic cells. Study the half-reactions carefully and try to understand them. Do not use half-reactions only as a means of balancing redox equations.

CHAPTER 17 PROBLEMS

17. 1a. What are the oxidation numbers of hydrogen and oxygen in hydrogen peroxide? Explain your choices.
Formula: H_2O_2
Answer: Oxidation numbers: H: +1 O: –1

Oxygen can be assumed to have an oxidation number of –2 except in those compounds in which oxygen is combined with a highly electronegative element and in complex unstable compounds. In this case, if both oxygens were –2 each, hydrogen would have an oxidation number of +2. This cannot be so, since hydrogen has only one electron. Hydrogen peroxide is highly unstable when concentrated.

17.1b. What are the oxidation numbers of fluorine and oxygen in OF_2? Explain your choices.

17.2a Apple cider is made by fermenting pressed apples. The fermentation produces a small amount of alcohol. Vinegar is made by continuing the fermentation of the cider. Alcohol reacts with oxygen to form, eventually, acetic acid. Assess the number of oxygens to determine whether this is an oxidation or a reduction. Confirm your result by showing the change in oxidation states of the alcohol and acid carbons. This unbalanced equation represents the reaction:

$$CH_3CH_2OH \longrightarrow CH_3COOH$$

Answer: The alcohol is oxidized, because the number of oxygen atoms increases.
(Assign O as –2, and H as +1 first. Then determine carbon charge needed to balance out.)
Oxidation states:

	Hydrogen	Oxygen	Carbon
In alcohol	6 @ +1	1 @ –2	2 @ –2
In acetic acid	4 @ +1	2 @ –2	2 @ 0

Carbon of alcohol has lost electrons (gone to the right on the redox number line). It has been oxidized.

17.2b. In the fermentation of sugars such as $C_6H_{10}O_5$, ethyl alcohol is produced by this reaction:

$$C_6H_{10}O_5 + H_2O \longrightarrow 2\ CH_3CH_2OH + 2\ CO_2\uparrow$$

According to oxygen content, is sugar oxidized or reduced? List the oxidation states of elements in the reactants and products to confirm your answer.

17.3a. Unsaturated fats are hydrogenated in the production of margarine. You have heard of unsaturated fats in nutrition. You will study them in Chapter 20. A molecule of an unsaturated fat contains double or triple bonds. Let's represent the unsaturated fat by showing a short segment of a molecule around a double bond. A vertical bar attached to a carbon represents a hydrogen bonded to the carbon. A horizontal bar attached to a carbon represents bonding to another carbon in a carbon chain.

```
 -C=C-
  |  |
  H  H
```

During hydrogenation the double bond opens and hydrogen atoms are added to either side of it.

```
  H  H
  |  |
 -C-C-
  |  |
  H  H
```

The carbons are now saturated. That means the maximum number of hydrogens are attached to the carbons. Is this therefore oxidation or reduction? What is the change in oxidation number of the two carbons?

Answer: Since this is an addition of hydrogen, it must be a reduction. When coupled by the double bond, the oxidation number of the carbons was −1. Each carbon was bonded to one hydrogen of oxidation state +1. [Carbon-to-carbon bonding does not change the oxidation number of carbon, just as the oxidation number of oxygen is not changed when it forms the molecule O_2.] After hydrogenation, the oxidation number of the carbons is –2 since two hydrogens are bonded to each carbon.

17.3b. During fermentation of starches, "fusel oils" are produced. These are poisons that are removed by distillation in making distilled spirits. The fusel oils are aldehydes and ketones. Is alcohol oxidized or reduced during reactions to form an aldehyde or ketone? Show the oxidation state changes of carbon in going from alcohol to an aldehyde. The unbalanced equation showing an alcohol giving an aldehyde product is:

$$CH_3CH_2OH \longrightarrow CH_3CHO$$

17.4a. Manganese can occupy many oxidation states from 0 to +7. Is the permanganate ion a reducing agent or an oxidizing agent?

Answer: The permanganate ion is MnO_4^-. Manganese is in the +7, its highest, oxidation state. It can't lose any more electrons, so it must act as an oxidant. The permanganate ion is an oxidizing agent.

17.4b. In the reaction (unbalanced equation) of the automotive lead-acid battery:

$$PbO_2 + Pb + H_2SO_4 \longrightarrow PbSO_4 + H_2O$$

What is being oxidized and what reduced? Name the oxidizing agent and reducing agent of the reaction. List the oxidation states of the elements involved.

17.5a. In the reaction as given by this unbalanced equation:

$$SnCl_2 + FeCl_3 \longrightarrow SnCl_4 + FeCl_2$$

What is being oxidized and what reduced? Name the oxidizing agent and reducing agent of the reaction. List the oxidation states of the elements that change oxidation state.

Answer: $SnCl_2$ is oxidized and is the reducing agent. $FeCl_3$ is reduced and is the oxidizing agent.

Oxidation states:	Tin	Iron
In $SnCl_2$	+2	
In $SnCl_4$	+4	
In $FeCl_3$		+3
In $FeCl_2$		+2

17.5b. Use this table of "electron grabbers" to answer this and other problems of this chapter. Will iron metal act as a reducing agent with permanganate?

Activity Series

easily reduced	MnO_4^-	$+ 8H^+$	$+ 5e^-$	$\rightleftharpoons$	Mn^{++}	$+ 4H_2O$	
↑	Cl_2		$+ 2e^-$	$\rightleftharpoons$	$2Cl^-$		
	O_2	$+ 4H^+$	$+ 4e^-$	$\rightleftharpoons$	$2H_2O$		
	Ag^+		$+ e^-$	$\rightleftharpoons$	Ag		
	Fe^{+++}		$+ e^-$	$\rightleftharpoons$	Fe^{++}		
	Cu^{++}		$+ 2e^-$	$\rightleftharpoons$	Cu		
	SO_4^-	$+ 4H^+$	$+ 2e^-$	$\rightleftharpoons$	SO_2	$+ 2H_2O$	
	$2H^+$		$+ 2e^-$	$\rightleftharpoons$	H_2		[acid]
	Fe^{++}		$+ 2e^-$	$\rightleftharpoons$	Fe		
	Zn^{++}		$+ 2e^-$	$\rightleftharpoons$	Zn		
	$2H_2O$		$+ 2e^-$	$\rightleftharpoons$	H_2	$+ 2OH^-$	[water]
	Al^{+++}		$+ 3e^-$	$\rightleftharpoons$	Al		
↓	Mg^{++}		$+ 2e^-$	$\rightleftharpoons$	Mg		
easily oxidized	Na^+		$+ e^-$	$\rightleftharpoons$	Na		

Ions are in water solution. All reactants and products, including gases, are at standard conditions. The table is ordered from the top down listing strong to weak electron grabbers. This table is similar to the Activity Series Table 10.1 in your text, which shows what metals will displace other metals from water solutions of their ions.

17.6a. Write the half equations for the reaction, if any, that occurs when metallic sodium is placed in water. Then add them to form the net ionic equation.

Answer: The half-reactions and net ionic reaction between sodium metal and water are:

$$2H_2O + 2e^- \rightleftharpoons H_2 + 2OH^-$$

$$2Na \rightleftharpoons 2Na^+ + 2e^-$$

Net Ionic: $$2Na + 2H_2O \rightleftharpoons H_2 + 2Na^+ + 2OH^-$$

17.6b. Write equations for the half-reactions and the net ionic equation for the reaction of chlorine and permanganate. What is reduced? What is oxidized?

17.7a. Describe the electroplating cell hookup and operation for plating zinc upon iron. Plating of zinc is performed in water solution of a soluble zinc compound. Refer to Figure 17.9 of your text.
Answer: The anode is zinc that is oxidized to Zn^{2+} ions in solution. These ions migrate toward the cathode. The cathode is the iron to be plated upon. Zn^{2+} ions are reduced at the cathode to metallic zinc Zn^0.
The positive side of the power supply is connected to the zinc anode, the negative side to the iron cathode. Iron is the cathode, since reduction occurs there. In the electrolytic cell, the zinc is reduced directly on the iron surface.

17.7b. Aluminum is produced by electrolysis of molten bauxite, an aluminum oxide ore, that contains Al^{+3}. Describe the electroplating cell hookup and operation of aluminum production. What goes where?

17.8a. In a galvanic cell using silver and copper, which is the cathode and which is the anode? Write the half-reactions and the net ionic reaction for this cell. Use the Activity Series Table. What direction do electrons flow in the external circuit?
Answer: The half-reactions and net ionic reaction between silver and copper are:

$$2\,Ag^+ + 2e^- \rightleftharpoons 2\,Ag$$

$$Cu \rightleftharpoons Cu^{++} + 2e^-$$

Net Ionic: $$2\,Ag^+ + Cu \rightleftharpoons 2\,Ag + Cu^{++}$$

The anode is the copper electrode, since copper is oxidized. Oxidation takes place at the anode. The cathode need not be silver, since silver ion is reduced to metallic silver at the cathode. Metallic silver would *plate out* on the cathode electrode material. The cathode material would have to be electrically conductive, of course. The oxidation of copper furnishes electrons to the external circuit. Electrons would flow from the copper anode through the external circuit to the cathode terminal. There Ag^+ picks them up.

17.8b. Write equations for the half-reactions and the net ionic reaction that occur when metallic zinc is placed in a solution of Cu^{2+}. What happens to the zinc? What happens to the copper?

Answers to "b" Problems

17.1b. O is +2, F is –1. Fluorine is highly electronegative, thus it takes the negative charge of –1, common for halogens. Fluorine is the only atom that can force oxygen to take a positive charge. The formula is written OF_2, not F_2O.

17.2b. Five oxygens in sugar are decreased to two in the alcohol products. Fewer O's means reduction. Sugar is reduced to the alcohol. The sugar carbons are at 0 charge; the alcohol carbons are at –2. [Some of the carbon from the sugar is oxidized to carbon dioxide, but we are to focus on the sugar/alcohol reaction.]

Oxidation states:	Hydrogen	Oxygen	Carbon
In sugar	10 @ +1	5 @ –2	6 @ 0
In water	2 @ +1	1 @ –2	
In alcohol	6 @ +1	1 @ –2	2 @ –2
In carbon dioxide		2 @ –2	1 @ +4

17.3b. The alcohol is oxidized. The number of hydrogens has gone from 6 down to 4. An increase in hydrogens is reduction, therefore a decrease in hydrogens is oxidation. The oxidation states of carbon are: –2 in alcohol; –1 in aldehyde, which is more positive, thus confirming that it is oxidation.

17.4b. PbO_2 is reduced and is the oxidizing agent. Metallic Pb is oxidized and is the reducing agent.

Oxidation states:	Lead	Sulfur	Hydrogen	Oxygen
In PbO_2	+4			–2
Pb metal	0			
$PbSO_4$	+2	+6		–2
H_2SO_4		+6	+1	–2
H_2O			+1	–2

17.5b. Yes. Iron is the reducing agent. Permanganate will be reduced. The half-reactions and net ionic reaction between iron and permanganate are:

Red: $2\,MnO_4^- + 16\,H^+ + 10\,e^- \rightleftharpoons 2\,Mn^{2+} + 8\,H_2O$

Ox: $5\,Fe \rightleftharpoons 5\,Fe^{2+} + 10\,e^-$

Net: $2\,MnO_4^- + 16\,H^+ + 5\,Fe \rightleftharpoons 2\,Mn^{2+} + 8\,H_2O + 5\,Fe^{2+}$

17.6b. Cathode: $5\,Cl_2 + 10\,e^- \rightleftharpoons 10\,Cl^-$

Anode: $2\,Mn^{2+} + 8\,H_2O \rightleftharpoons 2\,MnO_4^- + 16\,H^+ + 10\,e^-$

Net: $5\,Cl_2 + 2\,Mn^{2+} + 8\,H_2O \rightleftharpoons 10\,Cl^- + 2\,MnO_4^- + 16\,H^+$

Chlorine is reduced. Manganese is oxidized.

17.7b. The cathode is aluminum. The positive Al^{+3} ions are in the molten bauxite. They migrate toward the cathode and are reduced there. Metallic aluminum Al^0 is produced at the cathode of the electrolytic cell. Current (electrons) must be supplied to reduce the aluminum. The positive side of the power supply is connected to the anode, the negative side to the aluminum cathode. The anode is a conductive but inert electrode such as a graphite rod. Oxygen is produced at the anode.

17.8b. Zinc will become coated with copper metal. Zinc goes into solution as zinc ion; copper comes out of solution as copper metal. The half-reactions and net ionic reaction between zinc metal and copper ion are:

Anode: $Zn \rightleftharpoons Zn^{++} + 2e^-$

Cathode: $Cu^{++} + 2e^- \rightleftharpoons Cu$

Net Ionic: $Zn + Cu^{++} \rightleftharpoons Zn^{++} + Cu$

PRACTICE TEST I

1. What is the oxidation state of Cl in $Cu(ClO_4)_2$?
2. What is the individual charge on Cr in $MgCr_2O_7$?
3. What is the formal charge on B in BH_3?
4. The metabolism of food within the cell releases energy through the _______ of glucose.
5. Rapid, violent oxidation that radiates heat and light is called _______.

6a. What is reduced in the following reaction?

$$2K + I_2 \rightarrow 2KI$$

6b. What is the reducing agent?

7a. What is oxidized in the following reaction?

$$C_3H_8 + 5O_2 \rightarrow 3CO_2 + 4H_2O$$

7b. What is the oxidizing agent?

8. What is oxidized in the following reaction?

$$2SO_2 + O_2 \rightarrow 2SO_3$$

9. Is the reactant oxidized or reduced in the following reaction?

$$C_2H_5OH \rightarrow C_2H_4O + H_2$$

10. What is oxidized in the following unbalanced reaction, and by how many electrons?

$$H^+ + MnO_4^- + Cl^- \rightarrow Mn^{2+} + ClO_3^- + H_2O$$

11. What is the product when elemental nitrogen is completely oxidized by oxygen?
12. Which is the strongest reducing agent?

A) H_2 B) K C) K^+ D) F_2 E) Slim-Fast

13. Potassium permanganate makes a deep purple solution. It was once used as a foot bath for soldiers in the trenches, to kill bacteria and fungi. What type of agent is $KMnO_4$?

14. Liquid hydrogen, a reductant, is used as a rocket fuel. If you went to space and a shuttle dropped your rocket there with a load of $H_{2(l)}$, could you rocket home?

15. Which is an oxidizing agent?

A) aluminum

B) ammonia

C) hydrogen peroxide

D) sodium chloride

E) hydroquinone

16. Balance the following redox equation in acidic solution.

$$Fe^{2+} + IO_3^- \rightarrow Fe^{3+} + I^-$$

17. Balance the following redox equation in acidic solution.

$$MnO_4^- + C_2O_4^{2-} \rightarrow Mn^{2+} + 2CO_2$$

18. In a voltaic cell, what process takes place at the anode?

For questions 19-21, use the Activity Series (Table 10.1) to predict whether or not a cell using the following net reaction would produce an electric current; in other words, is the reaction spontaneous?

19. $Pb_{(s)} + CuSO_{4(aq)}$

20. $Zn_{(s)} + Au(NO_3)_{3(aq)} \rightarrow$

21. $Zn_{(s)} + MgSO_{4(aq)} \rightarrow$

22. Why are the two half-cell reactions in a voltaic cell separated?

PRACTICE TEST II

1. What is the individual charge on S in $AgHSO_3$?

2. What is the individual charge on C in BaC_2O_4?

3. What is the individual charge on C in CH_4?

4. Plants use sunlight energy and photosynthesis to store energy by the ________ of carbon dioxide.

5. The namesake and most common cause of oxidation is the chemical ________.

6a. What is oxidized in the following reaction?

$$Cu + O_2 \rightarrow CuO$$

6b. What is the oxidizing agent?

7a. What is reduced in the following reaction?

$$F_2 + S \rightarrow SF_6$$

7b. What is the oxidizing agent?

8a. What is reduced in the following reaction?

$$C_2H_2 + 2H_2 \rightarrow C_2H_6$$

8b. What is the reducing agent?

9. What is oxidized in the following reaction?

$$2FeCl_2 + Cl_2 \rightarrow 2FeCl_3$$

10. What is reduced in the following unbalanced reaction, and by how many electrons?

$NO_3^- + I^- \rightarrow NO + I_2$

11. The following reaction forms sodium hypochlorite, a component of laundry bleach.

$NaOH + Cl_2 \rightarrow NaCl + NaClO + H_2O$

Which statement about this reaction is true?

A) Only hydrogen has been oxidized.

B) Hydrogen and chlorine have been oxidized.

C) Chlorine has been both oxidized and reduced.

D) Oxygen is oxidizing other reactants, thus getting reduced itself.

E) This is not a redox reaction.

12. Which is most easily reduced?

A) F^- B) Ag^+ C) Na D) Zn E) your taxes

13. Give the name of a common household oxidizing agent used to remove stains.

14. Gangrene is caused by an infection of tissue with anaerobic bacteria, germs that can live only in the absence of air. However, if circulation can be maintained, these bacteria will die. What, exactly, causes their death?

15. Which is a reducing agent that is often used to smelt ores?

A) CO

B) bleach

C) O_2

D) potassium permanganate

E) Cl_2

16. Balance the following redox equation in acidic solution.

$NO_3^- + I^- \rightarrow NO + I_2$

17. Balance the redox equation below.

$\_H^+ + \_MnO_4^- + \_Cl^- \rightarrow \_Mn^{2+} + \_ClO_3^- + \_H_2O$

18. In a voltaic cell, reduction takes place at the ________.

19. An electrolytic cell to decompose water is constructed. What product collects at the anode, and what product collects at the cathode?

20. A student uses a good 9-volt battery connected to two electrodes. He places both electrodes in distilled water and waits for electrolysis to occur ... not! **Why** not?

21. Could gold from gold(III) nitrate solution be plated out spontaneously (without current) on a zinc base metal spoon dipped in it?

22. A voltaic cell with the following net reaction is constructed. Will it function (to produce current)?

$Pb_{(s)} + Hg(NO_3)_{2(aq)} \rightarrow Hg_{(l)} + Pb(NO_3)_{2(aq)}$

23. Can gold be oxidized by iodide?

24. Can sulfur be reduced by sodium fluoride?

Answers to Practice Test I

1. +7
2. +6
3. +3
4. oxidation
5. burning or combustion or fire

6a. I_2 6b. K

7a. C_3H_8 7b. O_2

8. SO_2
9. C_2H_5OH is oxidized, because it loses hydrogens.
10. Cl^-, by six electrons (to Cl^{+5})
11. N_2O_5
12. B
13. oxidizing agent
14. No! You would need liquid oxygen (the oxidant) too!
15. C
16. $6H^+ + 6Fe^{2+} + IO_3^- \rightarrow 6Fe^{3+} + I^- + 3H_2O$
17. $16H^+ + 2MnO_4^- + 5C_2O_4^{2-} \rightarrow 2Mn^{2+} + 10CO_2 + 8H_2O$
18. oxidation
19. yes
20. yes
21. no
22. If they were mixed, the electrons would be transferred directly between the reactants, and there would be no current of electrons traveling through the wire.

Answers to Practice Test II

1. +4
2. +3
3. −4
4. reduction
5. oxygen

6a. Cu 6b. O_2

7a. F_2 7b. F_2

8a. C_2H_2 8b. H_2

9. $FeCl_2$
10. (N of) NO_3^- by 3 electrons (N^{+5} to N^{+2})
11. C
12. B
13. chlorine bleach
14. The oxygen in the blood will kill (oxidize) them.
15. A
16. $8H^+ + 2NO_3^- + 6I^- \rightarrow 2NO + 3I_2 + 4H_2O$
17. The coefficients are: 18 6 5, 6 5 9. Be sure you try these yourself!
18. cathode
19. Oxygen gas appears at the anode and hydrogen gas at the cathode.
20. The student needs to add a salt (electrolyte) so the electron current can flow through the water.
21. Yes, the reaction is spontaneous.
22. No, it would not occur without a source of energy.
23. No. Iodide cannot take any more electrons, as it is already fully reduced.
24. No, sulfur cannot be reduced by sodium fluoride because Na^+ has no electrons to give, and fluoride as the most electronegative element will not give away electrons to sulfur.

CHAPTER 18

Fundamentals of Nuclear Chemistry

We've all heard about nuclear energy. Should we use nuclear energy? The use of nuclear energy has its problems, but so does any other endeavor of man. Coal burning in power plants has produced sulfur dioxide that seems to have contributed to acid rain and smog. If we scrub the SO_2 out, as we are doing, we are left with the production of carbon dioxide, which is contributing to global warmup. Coal mining destroys the natural landscape. Oil drilling may cause spills. Burning oil pollutes and wastes petroleum reactants. There are always problems.

Use of radioactive isotopes has been very beneficial in medicine. Some of the isotopes used are man-made, artificial. Some foods are now being irradiated for preservation. This process is a very efficient way of preserving foods. The foods last longer in storage, taste better, and have less potential of making us sick. The process kills insects and pathogenic bacteria and denatures enzymes that cause spoilage. Unlike chemical preservatives, irradiation leaves no residue.

The study of fundamental particles is the big thrust in modern theoretical physics. We are learning more about the structure of matter and energy and about the nature of the universe in which we live. The experimental work of particle physics requires massive accelerators, machines that 50 years ago were called "atom smashers". These machines have become extremely large and expensive, as more power was required of them to delve deeper into the fundamental particles and into the fundamental forces of nature. You might know that in 1994 our national Congress took away funding for an accelerator being built in Texas. It was called the SCC for superconducting collider. This was to be the largest accelerator in the world with an underground circular tunnel of several miles radius.

As we have learned more about the inner workings of the atom, we have developed many devices and techniques of great benefit to our society. Man is inquisitive. He wants to know about everything. Particle physics studies have no apparent immediate use, but if we did not pursue these kinds of studies as well as others, our progress would come to a standstill.

We virtually killed our nuclear power industry with too much safety regulation. Should we have done so? Should Congress have pulled the funding for the SCC? How much science should taxpayer money support? How much of anything should taxpayer money support? Try to learn enough about these various aspects of our modern society so that you can make intelligent judgments. Each has disadvantages, but each also has advantages. We have to learn to balance the benefits and risks, not emotionally but with scientific facts and studies.

SKILLS TO ENHANCE SUCCESS IN THIS CHAPTER

Realize that in this chapter we are in the realm of physics, not chemistry as such. It really doesn't matter. You should have no trouble with the material of this chapter. You can read elemental symbolism. Balancing nuclear equations is easier than balancing chemical ones. You know about electrical charge of the atomic particles: The electron is negative, the proton positive, the neutron neutral.

Here the isotope is king, not the whole element. In nuclear reactions, one does not speak of uranium, but of uranium-235 or some other number. You already know about isotopes. One new idea is half life. This requires use of exponents. Learn how to use your calculator's X^y button. Keep your bookkeeping skills up, as well as your microscopic visualization ability. Then you will be OK in this chapter.

CHAPTER 18 PROBLEMS

18. 1a. What isotope is produced when strontium–90 decays by emission of a beta particle? What element is produced?

Given: $^{90}_{38}Sr$

Need: Product of beta emission

Connecting Information:

Beta particle: electron, $^{0}_{1-}e$

Symbolism

Pre-superscript: nuclear mass

Pre-subscript: nuclear charge or atomic number; determines element

Principles involved

Conservation of mass

Conservation of charge

1. Write equation using $^{a}_{b}X$ as symbol for product isotope.

$$^{90}_{38}Sr \longrightarrow \, ^{a}_{b}X + \, ^{0}_{1-}e$$

2. Determine values of a and b.

$90 = a + 0$ so, $a = 90$

$38 = b + (-1)$ so, $b = 39$

Find identity of X: nuclear charge = 39 = atomic number 39

Isotope of yttrium, Y–90

3. Write equation.

$$^{90}_{38}Sr \longrightarrow \, ^{90}_{39}Y + \, ^{0}_{1-}e$$

The pre-superscripts and the pre-subscripts are separately balanced across the equation. Check the sums.

18. 1b. What isotope is produced when silver–104 decays by emission of a beta particle? What element is produced? Write the balanced nuclear equation.

18.1 a. If an isotope of curium emits an alpha particle, an isotope of what element is produced?

Given: Cm: atomic number 96

Need: Element produced by alpha emission

Connecting Information:

Alpha particle: helium nucleus, atomic number 2

Atomic number is reduced by 2 after alpha emission. ($96 \longrightarrow 94$)

Element produced: Pu, plutonium, atomic number 94.

18.1 b. If fluorine decays by beta emission, what element is produced?

18.2 a. Technicium–99, with a half-life of 6 hr, is used to detect brain tumors. If 1 mg of this isotope is injected into a patient, how much remains in the patient after 24 hr? Assume no elimination of the isotope by the patient.

Given: Tc–99 radioisotope: 1 mg initial quantity

Need: Remaining amount after 24 hr

Connecting Information:

Half-life: 6 hr

$$\text{Number of half-lives: } \frac{24 \text{ hr}}{6 \text{ hr}} = 4$$

$$\text{Divisor: } 2^4 = 16$$

$$1 \text{ g} \times \frac{1}{16} = 0.0625 \text{ g remaining.}$$

18.2b. Chromium–51 is used in the determination of red blood cell quantity. Its half-life is 28 days. If the amount of chromium–51 in a patient is 12.5% of the original amount injected, how long ago was the tracer injected into the patient? Assume no elimination of the isotope by the patient.

18.3a A kilogram of matter is equivalent to how much energy in joules?
Given: 1 kg, any matter
Need: energy of mass
Connecting Information:
Einstein equivalency: $E = mc^2$
Energy: $1\,J = 1\,kg \times \frac{1\,m^2}{1\,sec^2}$
Velocity of light: $c = 2.9979 \times 10^8 \frac{m}{sec}$

$$\text{Energy in Joules: } E = mc^2 = 1\,kg \times \left[2.9979 \times 10^8 \frac{m}{sec}\right]^2$$
$$= 8.9874 \times 10^{16}\,J = 8.9874 \times 10^{13}\,kJ$$

18.3b. Express this energy in calories. How many rads is that?
1 rad = 100 ergs absorbed
2.39×10^{-8} cal/erg
4.184 J/cal
1 x 10–7 J/erg

18.4a. Name a device that measures the instantaneous amount of radiation impacting at just that moment.
Answer: The Geiger Counter
18.4b. Name a radiation detector that is used to detect cumulative radiation totals.

18.5a. What type of source does most of our radiation exposure come from?
Answer: Natural sources, not man-made sources, are the primary cause for background radiation exposure.
18.5b. Which natural source is the largest contributor to radiation exposure?

18.6a When molybdenum–98 absorbs a neutron, it emits an electron. What isotope is produced during this reaction?
Given: $^{98}_{42}Mo$; emits electron upon neutron absorption
Need: Neutron absorption products
Connecting Information:
Neutron: $^{1}_{0}n$ Electron: $^{0}_{1-}e$

$$^{98}_{42}Mo + {}^{1}_{0}n \longrightarrow {}^{99}_{43}Tc + {}^{0}_{1-}e$$

Technicium–99 isotope produced.
18.6b. Astatine-211, not found in nature, was synthesized by alpha bombardment of bismuth–209. What other products were observed during this experiment?

18.7a To synthetically produce radioactive carbon, what element would you pick as the "target" element and with what radiation would you bombard it?
Answer: Carbon–12 is a very stable isotope. It is the common isotope present in organic compounds and in our own bodies. It is also the isotope selected as a base for determination of atomic masses. The carbon–12 nucleus contains six neutrons. A carbon nucleus with more than six neutrons would likely be unstable, radioactive. Carbon–14 is used in radioactive dating. Putting two more neutrons in the carbon nucleus would produce carbon–14. To bombard the carbon nucleus itself with neutrons would not produce carbon–14. Upon bombardment, another particle will be emitted. The best bet is to pick a target nucleus that already has the required number of neutrons. Such a nucleus would be nitrogen–14. Bombard it with neutrons. The reaction is:

$$^{14}_{7}N + ^{1}_{0}n \longrightarrow ^{14}_{6}C + ^{1}_{1}H$$

One could conceive of bombarding a lower number element with alpha particles. But an alpha particle and a nucleus of an element of number lower than carbon are about equal in size. Fusion of light nuclei requires extremely high temperatures and pressures to occur.

18.7b. What element could be used as the target element to produce californium? With what would you bombard?

18.8a A sample of ancient wood was analyzed and found to contain 6.25% of assumed original C−14. How old is the wood?

Given: C−14 abundance: 6.25% of original

Need: Age of wood sample

Connecting Information:

Half-life C−14: 5730 yr

Solution 1: Countdown method

Tabulate years from present and C−14 remaining at end of each half-life time span.

Half-life	Years Elapsed	Remaining C−14, %
1st	5730	50 %
2nd	11460	25 %
3rd	17190	12.5 %
4th	22920	6.25 %

Wood is approximately 23,000 years old.

Solution 2: Exponential calculation

1. Determine number of half-lives that have elapsed.

$$\frac{1}{2^n} = 0.0625 = \frac{1}{16}$$

$$2^n = 16$$

$$n = 4 \text{ half-lives}$$

2. Determine age of wood.

$$5730 \text{ yr/half-life} \times 4 \text{ half-lives} = 22920 \text{ yr}$$

18.8b. Iodine−131 is used for detection of abnormalities in glands of the human body. What fraction of this isotope would remain in a patient a month after injection? The half-life is eight days.

18.9a Explain the meaning of "chain reaction". Use the reaction below as an example.

$$^{235}_{92}U + ^{1}_{0}n \longrightarrow ^{139}_{56}Ba + ^{94}_{36}Lr + 3^{1}_{0}n$$

Answer: A chain reaction is one that increases in intensity with time of reaction, and propagates itself. The reaction of a small amount of material releases energy which, being absorbed by nearby material, initiates reaction of more of the nearby material. This action continues and is accelerated in this manner. The energy sustaining a nuclear chain reaction is a particle. Fission of a nucleus is initiated by absorption of a neutron. The reaction given above is a nuclear chain reaction described by chemists Fritz Strassmann and Otto Hahn and physicist Lise Meitner. Three neutrons are emitted. Each of these neutrons can initiate fission in another uranium−235 nucleus.

18.9b. Why is a "critical mass" required for chain fission?

18.10a. How are the control rods in fission reactors used for production of electrical energy?

Answer: The control rods are moved in and out between cylinders of the "fuel" within the reactor to control the rate of overall reaction. The rods absorb neutrons and slow the overall reaction of the total mass of fissionable material in the reactor.

18.10b. What is the purpose of water in a typical U.S. nuclear plant, and why is it kept under high pressure?

18.11a. How much energy in kJ is available if the mass of 1 gram of hydrogen were completely converted to energy? (Recall problem 18.3a.) The actual output from fusion of one gram of hydrogen in a fusion reaction is as much energy as that released during burning of 20 tons of coal (6×10^8 kJ). What percent of the hydrogen mass is actually converted to energy in fusion? (That is the efficiency of fusion generators.)

Given: 1 g hydrogen fusing to helium
1.00 kg of mass could produce 8.9874×10^{13} kJ (from problem 18.3a)
Actual energy output is 6×10^{8} kJ.

Need: Energy from conversion of 1.00 g of mass to energy
% actual out of total available

Calculate:

1. Energy from 1 gram of hydrogen if all were converted:

$$\frac{8.9874 \times 10^{13}\ \text{kJ}}{1\ \text{kg mass}} \times \frac{1\ \text{kg}}{1000\ \text{g}} \times 1.00\ \text{g} = 8.9874 \times 10^{10}\ \text{kJ} = 9 \times 10^{10}\ \text{kJ}$$

2. % actual out of total available:

$$\frac{6 \times 10^{8}\ \text{kJ}}{9 \times 10^{10}\ \text{kJ}} \times 100\% = 0.66\%\ \text{conversion or } 0.7\%$$

18.11b. What happened to the other 99.3% of the energy of hydrogen?

Answers to "b" Problems

18.1b. Cadmium–104 produced. $^{104}_{47}\text{Ag} \longrightarrow {}^{104}_{48}\text{Cd} + {}^{0}_{1-}\text{e}$

18.1b′. Neon, Ne, element number 10

18.2b. 12.5% = one-eighth of 100%. 1/8 = $1/2^3$. Three half-lives @ 28 days = 84 days ago.

18.3b. 2.148×10^{16} cal ÷ 2.39×10^{-8} cal/erg = 8.99×10^{23} erg ÷ 100ergs/rad = 8.99×10^{21} rads

18.4b. A film badge

18.5b. Radon is the principle source of natural radiation.

18.6b. Two neutrons: $^{209}_{83}\text{Bi} + {}^{4}_{2}\text{He} \longrightarrow {}^{211}_{85}\text{At} + 2\,{}^{1}_{0}\text{n}$

18.7b. Pick curium. Bombard it with alpha particles. Heavy nuclei will absorb helium nuclei. The reaction is:

$$^{242}_{96}\text{Cm} + {}^{4}_{2}\text{He} \longrightarrow {}^{245}_{98}\text{Cf} + {}^{1}_{0}\text{n}$$

18.8b. $1/2^4$ is 1/16 left, or about 6.3%. A month is close to 32 days, 4 half-lives of iodine−131.

18.9b. If an insufficient number of tightly packed uranium atoms does not exist, not all the emitted neutrons from one fission will strike and be absorbed by other uranium nuclei. These neutrons would instead be lost to the reaction mass, and would not strike other nuclei to initiate further fission. The reaction will not propagate. It may even die down.

18.10b. The water is used both as a moderator and as a coolant. It is kept under pressure so that it can get hotter than 100°C without boiling.

PRACTICE TEST I

1. Alpha radiation consists of
   A) high-energy electrons.
   B) helium nuclei.
   C) X rays.
   D) protons.
   E) high-energy neutrons.

2. Gamma radiation
   A) consists of protons.
   B) is high energy electrons.
   C) consists of neutrons.
   D) consists of particles having neither mass nor charge.
   E) turns Doc Bruce Banner into the Hulk.

3. Transmutation is
   A) caused by chemical reaction.
   B) genetic damage caused by radiation.
   C) one element changing into another.
   D) the mutation of the electron configuration.
   E) a new wave group.

4. What percent of radioactive isotopes remains after three half-lives?

5. The half-life of Einsteinium-243 is 20 seconds. If you are given 1 gram of this isotope, how much remains after one hour has elapsed?

6. Which is equal to a curie, Ci?
   A) 1,000,000 rads
   B) 1,000,000 roentgens
   C) 3.7×10^{10} disintegrations/second
   D) 1 kcal
   E) 100 ergs of energy absorbed by tissue

7. A Geiger counter detects
   A) phosphors.
   B) ionizing radiation.
   C) electrons.
   D) tiny flashes of light.
   E) little clicks or noises.

8. Which type of radiation meter detects tiny flashes of light produced by radiation impact?

9. Which form of naturally occurring radiation has the least penetrating power?

10. High levels of radon-222 in the home
   A) are harmless.
   B) can cause lung cancer.
   C) are difficult to detect.
   D) cause immediate coma and death.
   E) cannot be corrected; destroy the house.

11. Most (82%) of our exposure to radiation
   A) is naturally-occurring radiation.
   B) is from X rays.
   C) comes from reactions within our bodies.
   D) comes from nuclear power plants.
   E) comes from reading Chapter 18 of Burns.

12. One way of producing artificial elements is
   A) by fission.
   B) UV irradiation.
   C) bombardment of target nuclei.
   D) through chemical reactions.
   E) with alchemy.

13. A beneficial use of radioisotopes is in
    A) microwave ovens.
    B) gamma radiation to preserve foods.
    C) AM/FM waves.
    D) manufacture of silverware.
    E) cell phones.

14. Radioisotopes have been used to
    A) power nuclear plants.
    B) power pacemakers.
    C) date artifacts.
    D) trace biological substances in living cells.
    E) all of the above.

15. "Atom splitting" is
    A) nuclear fission.
    B) caused by X rays.
    C) nuclear fusion.
    D) cutting an atom in half.
    E) chemical decomposition.

16. The control rods in the first uranium-235 chain reaction attempts by Fermi's team in Chicago were made of ___________, which absorbs neutrons.

17. Why is it so difficult to use fusion instead of fission as a power source?

18–22. Complete the following nuclear reactions using the form ^{A}X where A is the mass number and X is the element's symbol. Then classify what type of nuclear reaction it is. [Hint: You may wish to write in Z, the atomic number, as a subscript under superscript A.]

18. $^{222}Rn \rightarrow {}^{218}Po +$ ______.

19. $^{234}Th \rightarrow {}^{234}Pa +$ ______.

20. $^{14}N + {}^{1}n \rightarrow$ ______ $+ {}^{1}H$

21. $^{235}U + {}^{1}n \rightarrow {}^{142}Ba + {}^{91}Kr + 3$______.

22. $^{2}H + {}^{3}H \rightarrow$ ______ $+ {}^{1}n$

23. Emission of a positron can be thought of as converting
    A) an electron into a neutron.
    B) a neutron into a proton.
    C) an electron into a proton.
    D) a proton into an electron.
    E) a proton into a neutron.

24. Electron capture has the same result as
    A) positron emission.
    B) beta decay.
    C) neutron capture.
    D) alpha decay.
    E) gamma decay.

25. What is the last isotope in this series?

    ^{238}U alpha decay $\rightarrow$ ______ beta decay $\rightarrow$ ______ beta decay $\rightarrow$ ______ alpha decay $\rightarrow$?

PRACTICE TEST II

1. Beta radiation consists of
   A) high-energy electrons.
   B) helium nuclei.
   C) X rays.
   D) protons.
   E) high-energy neutrons.
2. X rays
   A) are produced by electron transitions.
   B) are produced by nuclear transitions.
   C) consist of neutrons.
   D) have a mass of 1 amu.
   E) have a +1 charge.

3. Natural radioactivity occurs in about 80 out of 350 naturally-occurring isotopes because they have
   A) high-energy electrons. C) heavy neutrons.
   B) unstable nuclei. D) been excited. E) been ionized.
4. What fraction of radioactive isotopes remains after five half-lives?
5. Certain malignant tumors are slowed by gamma radiation from cobalt-57 administered intravenously. The half-life of ^{57}Co is nine months. What percent of the ^{57}Co remains in the body after three years?
6. Which is equal to a rad?
   A) 1,000,000 Grays
   B) 1,000,000 roentgens
   C) 3.7×10^{10} disintegrations/second
   D) 1 kcal
   E) 100 ergs of energy absorbed by tissue

7. Name a radiation detector used for measuring cumulative absorbed dose.
8. X rays used for diagnostic purposes average 1 rad per year. The x-ray technician leaves the room to avoid exposure to this radiation. Should <u>you</u> avoid X ray exposure as a routine part of your health care?
9. What material provides an excellent barrier to radiation? (Keep your kryptonite in a box made of this.)
10. People who live at high altitudes
    A) are exposed to more uranium radioactivity.
    B) are less susceptible to alpha radiation.
    C) receive more cosmic radiation, man.
    D) receive more radon-222 exposure.
    E) should check in to a rehab clinic.

11. Most of our exposure to "natural" radiation comes from
    A) radon-222.
    B) uranium-238.
    C) reactions within our bodies.
    D) ground water.
    E) cigarette smoke (14.6 rem/yr).

12. Artificial transmutation can be accomplished by
    A) X rays. C) bombardment of target nuclei.
    B) UV irradiation. D) chemical reactions. E) plastic surgery.
13. Beneficial uses of radioisotopes include
    A) PET scans. B) food preservation. C) nuclear medicine. D) all of these.

14. Radioactive dating
    A) is done with carbon-12.
    B) takes advantage of a knowledge of isotopic half-lives.
    C) is quite unreliable.
    D) may be hazardous to your health. Choose a nonradioactive partner.

15. The source of the sun's energy is
    A) nuclear fission.
    B) X rays.
    C) chemical decomposition.
    D) combustion.
    E) nuclear fusion.

16. In the 1940s, to moderate the chain reaction of fissionable uranium-235, the Germans tried ________ while the Americans used ________ to slow down the neutrons.

17. Why did the Russian Chernobyl accident in 1986 cause more widespread radiation sickness than the American 1979 Three Mile Island accident?

18–22. Complete the following nuclear reactions using the form ^{A}X where A is the mass number and X is the element's symbol. Then classify what type of nuclear reaction it is. [Hint: You may wish to write in Z, the atomic number, as a subscript under A. Using this notation, the symbol $^{0}e^{-}$ represents a beta particle.]

18. $^{239}Pu + {}^{1}n \rightarrow {}^{0}e^{-} +$ ______.

19. $^{235}U \rightarrow {}^{231}Th +$ ______.

20. $^{90}Sr \rightarrow$ ______ $+ {}^{0}e^{-}$

21. $^{214}Bi \rightarrow {}^{0}e^{-} +$ ______.

22. $^{235}U + {}^{1}n \rightarrow {}^{99}Zr +$ ______ $+ 2\,{}^{1}n$

23. What product results when ^{20}Na undergoes positron emission?

24. Emission of a β particle can be thought of as converting
    A) an electron into a neutron.
    B) a neutron into a proton.
    C) an electron into a proton.
    D) a proton into an electron.
    E) a proton into a neutron.

25. What is the product when the ^{7}Be nucleus captures an electron from its own electron cloud?

26. The critical mass of uranium-235 is about the size of a/an
    A) ant.
    B) baseball.
    C) crib.
    D) davenport.
    E) elephant.

27. Radiation we receive comes from

    A) outer space.
    B) the earth's crust.
    C) radioactive isotopes in our own bodies.
    D) X rays for medical and dental purposes.
    E) all of the above.

Answers to Practice Test I

1. B
2. D
3. C
4. 12.5%
5. 6.5×10^{-55} g (180 half-lives)
6. C
7. B
8. a scintillation counter
9. alpha radiation
10. B
11. A
12. C
13. B
14. E
15. A
16. cadmium
17. Fusion involves extremely high temperatures.
18. ^{4}He; alpha decay
19. ^{0}e ; beta decay
20. ^{14}C; neutron capture
21. ^{1}n (three neutrons)
22. ^{4}He; nuclear fusion
23. E
24. A
25. $^{238}_{92}U$ alpha decay → $^{234}_{90}Th$ beta decay → $^{234}_{91}Pa$ beta decay → $^{234}_{92}U$ alpha decay → $^{230}_{90}Th$

Answers to Practice Test II

1. A
2. A
3. B
4. 1/32
5. 6.25%
6. E
7. A film badge
8. The benefits far outweigh harmful effects. (Technicians leave to avoid unnecessary long-term chronic exposure.)
9. lead
10. C 11. A 12. C 13. D 14. B 15. E
16. heavy water . . . graphite rods
17. The Chernobyl plant had no containment building.
18. ^{240}Am; neutron capture (or bombardment)
19. ^{4}He; alpha decay
20. ^{90}Y; beta decay
21. ^{214}Po; beta decay
22. ^{135}Te; fission
23. ^{20}Ne
24. B
25. ^{7}Li
26. B
27. E

CHAPTER 19

Organic Chemistry

Organic chemistry was originally considered to be a study of compounds that could be produced only during life processes, during metabolism in plants and animals. In the late 1800s, urea (a component of urine) was synthetically produced in a laboratory, and soon other metabolic compounds followed. Organic chemistry is now described as the chemistry of carbon compounds, whether from natural or laboratory sources. There is some overlap between inorganic and organic chemistries. The distinction is ultimately arbitrary, and is somewhat flexible.

SKILLS TO ENHANCE SUCCESS IN THIS CHAPTER

Organic chemistry seems overwhelming at first. So many compounds! Don't let it throw you. You have sufficient knowledge at this time to achieve a basic understanding of any chemistry. You know what the chemical bond is and how to join atoms into molecules. In organic chemistry, you join carbon atoms in long chains. Atoms attached to the carbon chain are mostly hydrogen, but other chains (sidechains) can also replace the hydrogens. Sometimes other elements join the chain. You should review molecular structure and Lewis electron-dot diagrams. Memorize that carbon prefers four bonds, nitrogen three bonds, oxygen two, and hydrogen one bond. That will help you draw structures quickly. You may be in for one surprise. Carbon can join in rings. Intramolecular bonding in these rings is covalent, but of a highly modified form of covalency. We will not delve into this bonding.

We always classify to give order and pattern to our world. In organic chemistry, to make sense of the large number of possible compounds, we classify them into groups or series of *homologs* (compounds related by structure). Each series has a common portion, a *functional group*, which is an assemblage of atoms attached to the main carbon chain or ring. This functional group determines the major properties common to the compounds within the series.

CHAPTER 19 PROBLEMS

19.1a. You are given an unknown material and asked to classify it as organic or inorganic. Visual examination reveals crystals. Measurements of physical properties result in a determination that the melting point is extremely high and that the compound is soluble in water. Tests of chemical reactivity show that it is not flammable. What kind of compound is it?
Answer: This is an inorganic compound as evidenced by solubility and high melting point. It is probably pure, as evidenced by crystalline structure.

19.1b. Your next unknown is a solid at STP that has a molar mass of 10,000 grams. Is the material organic or inorganic?

19.2a. A certain alkane has a molar mass of 30 grams. What is the substance?
Given: Alkane, molar mass = 30 g
Need: Identification of alkane
Connecting Information:
Alkane: consists of only carbon and hydrogen, carbons singly bonded

1. Write the general formula for an alkane.

$$C_nH_{2n+2}$$

The general formula immediately shows the pattern of the alkane homologous series.

2. Calculate n, using atomic masses of carbon and hydrogen.

$$12n + 1(2n+2) = 30$$

$$14n + 2 = 30$$

$14n = 28$ therefore, $n = 2$

3. Write the formula for the substance:

$$C_2H_{4+2} = C_2H_6$$

4. Identify the substance:

It is ethane.

19.2b. An alkane has a molar mass of 58. Identify the alkane.

19.3a. Show all the isomers of butane, sketching only the backbones of the compounds.
Answer: These are the isomers of butane:

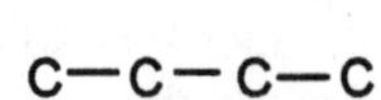

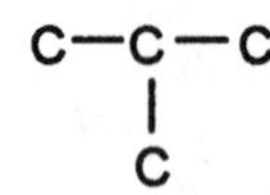

19.3b. Write structural formulas for these two isomers.

19.4a. Name the above isomers of butane using two different styles.
Answer: The IUPAC names are *n*-butane and 2-methylpropane. The older names are butane and isobutane. The *n*- prefix stands for normal, or straight chain. It can be left out if no confusion would result.

19.4b. Name the isomers of pentane using IUPAC rules.

19.5a. Show all the isomers of butene, sketching only the backbones of the molecules.
Answer: The isomers of butene are:

C—C—C=C C—C=C—C C—C=C (with C attached below the middle C)

19.5b. Name the isomers of butene. How are the carbons numbered in 1-butene? How are they numbered in 2-butene?

19.6a. Name this compound:

(benzene ring)—Br

Answer: The compound is bromobenzene.

19.6b. Give another name for 1,3-dichlorobenzene.

19.7a. Explain why the following names for two alcohols are incorrect:
2-ethyl-2-butanol 4-ethyl-2-pentanol.
Give the correct names for these compounds.
Answer: These would be the structural formulas, written directly from the names as given:

$H_3C—C(OH)(C_2H_5)—C—CH_3$ $H_3C—C(OH)—C—C(C_2H_5)—CH_3$

The longest chain containing the alcohol group, –OH, is not 4 but 5 carbons. The compound is a pentanol. The correct name is 3-methyl-3-pentanol

The longest chain containing the alcohol group, –OH, is not 5 but 6 carbons. The compound is a hexanol. The correct name is 4-methyl-2-hexanol.

You understand that hydrogens are attached where not directly shown in this formula.

19.7b. Is 2-methyl-1-propanol a primary, secondary, or tertiary alcohol? Give two other, older, names for this compound.

19.8a You and fellow students are doing homework together. One of the homework problems is the naming of a seemingly simple aromatic compound. But you disagree on the name. Here are the names that you and your fellows come up with:

- 3-chloro-hydroxybenzene
- 3-hydroxy-chlorobenzene
- 3-chloro-phenyl alcohol
- 3-chloro-1-hydroxybenzene
- 3-hydroxy-1-chlorobenzene
- 3-chloro-benzyl alcohol

Which of these names is correct? (Hint: Try drawing the structural formula for the compound from each of these names.)

Answer: None of the names are actually correct. The compound's actual name is 3-chlorophenol. Standards of nomenclature have to be agreed upon within an international committee for communication purposes. Groups of atoms that can be considered parent compounds are given special names. In naming a compound, one must be careful to select as a parent name one that is meaningful to a reader. For the compound of this problem, phenol, a name of long standing, should be the parent name within the name of the compound. An alternative correct name would be *meta*-chlorophenol, or *m*-chlorophenol.

19.8b. Your same group of students comes up with two names for another compound: 3-chloro-5-chloro-1-hydroxybenzene and 3-chloro-1-hydroxy-5-chlorobenzene. Are these names correct? Give the accepted name for this compound.

19.9a. Name $(CH_3CH_2CH_2)_2O$ the "easy" way and the IUPAC way.

Answer: This is an ether, called dipropylether or propoxypropane.

19.9b. What is the structural formula of methoxypentane?

19.10a Name this compound:

$$H_3C{-}CHCl{-}CH_2{-}CO{-}CH_3$$

Answer: The compound is 4-chloro-2-pentanone.

This compound could also be called 2-chloro-4-pentanone, but the 2-pentanone name is preferable, since the functional group rules. The idea is to number the carbons choosing the end that results in the lowest numbers possible for the side chains. If two equally low-numbered choices result, then preference should be given to the main functional group for the lower number.

19.10b. Write the condensed formula for 4-bromo-2-chloro-3-pentanone.

19.11a Name this compound:

$$H_3C{-}CH(CH_3){-}CH_2{-}COOH$$

The symbolism $-CH(CH_3)-$ represents a methyl side chain attached to the C of the CH.

Answer: The compound is 3-methylbutanoic acid.

Carbon number one is the carboxylic carbon of $-COOH$.

19.11b. Name this compound:

$$H_3C{-}CH(C_2H_5){-}CH(CH_3){-}COOH$$

How should the formula be written?

19.12a Name this compound:

$$H_3C{-}(CH_2)_3{-}NH{-}CH_3$$

Answer: The name of the compound is butylmethylamine.

In naming the alkyl groups attached to the N of an amine, the largest group is named first.

19.12b. Name this compound:

$$H_3C{-}CH(CH_3){-}CH_2{-}NH{-}CH_3$$

Answers to "b" Problems

19.1b. The material is organic. No inorganic compound has a molar mass of this magnitude.
19.2b. butane
19.3b. $CH_3CH_2CH_2CH_3$ and $CH_3CH(CH_3)_2$
19.4b. The three isomers of pentane are *n*-pentane, 2-methylbutane, and 2,2-dimethylpropane.
19.5b. 1-butene, 2-butene, and 2-methyl-1-propene are the isomers of butene. In 1-butene, number 1 carbon is the end carbon that is doubly bonded, since preference for lower numbering is given to the functional group. If we numbered from the other end, it would come out as 4-butene, but the lower number is preferred. For the other isomer, the double bond in the middle is the same distance from each end, so it wouldn't matter (lacking other substituents) which end we started numbering at. Either way, it would come out as 2-butene.
19.6b. *m*-dichlorobenzene.
19.7b. It is a primary alcohol. The hydroxyl group is on a primary carbon, a carbon bonded to only one other carbon. Older names are isobutanol and isobutyl alcohol.
19.8b. Neither of these names is correct. The name of the compound is: 3,5-dichlorophenol.
19.9b. $CH_3OC_5H_{11}$
19.10b. $H_3C{-}CHCl{-}CO{-}CHBr{-}CH_3$
19.11b. 2,3-dimethylpentanoic acid. The formula should be written $H_3C{-}CH_2{-}CH(CH_3){-}CH(CH_3){-}COOH$
19.12b. Isobutylmethylamine. It could also be named 2-methylpropyl-methylamine.

PRACTICE TEST I

1. In general, what makes a substance "organic", chemically speaking?
2. Which of these is an alkane?
   A) C_6H_6 B) C_6H_{10} C) C_6H_{12} D) C_6H_{14} E) C_6H_1
3. The alkane $C_{12}H_{26}$ is called _______ .
4. The carbon atom normally has how many bonds?
5. The isopropyl group can be written as _______ .
6. Two hydrocarbons with the same chemical formula but a different structure are called _______ .
7. Alkenes contain a ______ ______ between two of their carbon atoms.

NAME THE FOLLOWING HYDROCARBONS.

8. $CH_3CH{=}CH_2$

9.
```
CH3CH2CHCH3
        |
       CH3
```

10.
```
CH3CHCH2CH2CHCH3
   |        |
  CH3      C2H5
```

11.
```
CH3(CH2)5 CH(CH2)3CH(CH2)2CH3
          |       |
         CH3   CH3CHCH3
```

12.
```
CH3CHCH2CH2CHCH2CHCH3
   |        |    |
  CH3     C2H5  CH3
```

13.
```
         CH2CH2CH2CH3
         |
CH3CH2CHC=CHCH3
      |
      CH2
      |
      CHCH2CH3
      |
      CH3
```

14. What is polymerization?

15. Hydration of butene would yield what product?

16. Polystyrene, when mixed with tiny air bubbles and extruded, makes the familiar white material known as ________.

17. Aromatic compounds contain what group?

18. Name

A) *cis*-dichlorobenzene
B) *m*-dichlorobenzene
C) *p*-dichlorobenzene
D) *o*-dichlorobenzene
E) dichlorobenzene

19. Give the structural formula for diethyl ether.

20. Give the structural formula for formaldehyde.

21. Give the structural formula for 3-heptanone.

22. Give the condensed structural formula for octanoic acid.

23. Give the condensed structural formula for ethylamine.

24. The structure
```
R—C—N—
  ||  |
  O
```
represents what functional group?

25. Give two *specific* examples of a polymeric material.

26. Oils and flavorings often contain what functional group?

27. List the two types of polyethylene in order of increasing rigidity.

PRACTICE TEST II

1. Which of the following are organic?
   A) lipids
   B) carbohydrates
   C) alcohols
   D) plastics
   E) all of the above

2. What is the formula for nonane?

3. There are eighteen different isomeric forms of octane due to ________ of the carbon "backbone".

4. Melting and boiling points of alkanes ________ with an increasing number of carbons.

5. How does one select the root name for a branched chain of carbons?

6. What family of hydrocarbons contains a triple bond between two carbons?

7. What is a diene?

WRITE THE CONDENSED STRUCTURAL FORMULA OF EACH OF THE FOLLOWING.

8. 1-fluoro-2-butene
9. 3-methyl-6-*tert*-butylnonane
10. 4-ethyl-5,7,7-trimethylundecane
11. 4,4-diisopropyl-1-octene
12. *trans*-3-hexene

13. What is the name of the product one gets when bromine is added to the double bond of 1-pentene?
14. What is the general name for hydrocarbons to which the NH_3 group is attached?
15. Ethylene
    A) is an alkane.
    B) ripens tomatoes.
    C) is a polymer.
    D) is a saturated hydrocarbon.
    E) is aromatic.
16. Distinguish aliphatic from aromatic compounds.
17. Glycerol, a triol, has what condensed structural formula?

MATCH THE FUNCTIONAL GROUP SHOWN WITH ITS NAME.

18. —C(=O)—N(—)— (C double-bonded to O below; N with a bond below)

19. —C(|)(|)—O—C(|)(|)—

20. —C(H)=O

21. —C(|)(|)—N(|)—

22. —C(|)(|)—C(=O)—C(|)(|)—

23. —C(=O)—O—C(|)(|)—

23. —C(=O)—O—C(|)—

24. —C(=O)—OH

A) ester
B) ether
C) ketone
D) aldehyde
E) carboxylic acid
F) amine
G) amide

25. Amide linkages are found in
    A) nylon.
    B) proteins.
    C) stearic acids.
    D) dienes.
    E) polyethylene.

MATCH THE TYPES OF COMPOUNDS WITH THEIR ODORS.

26. esters	A) fishy, decaying flesh
27. carboxylic acids	B) pungent, cool, like charcoal lighter fluid
28. amines	C) sweet and fruity
29. alkanes	D) foul, rancid, cheesy

Answers to Practice Test I

1. If it contains carbon, it's organic.
2. D
3. dodecane
4. four
5. —C_3H_7 or $CH_3\overset{|}{C}HCH_3$
6. isomers
7. double bond
8. propene
9. 2-methylbutane
10. 2,5-dimethylheptane
11. 4-isopropyl-8-methyltetradecane
12. 2,7-dimethyl-4-ethyloctane (The number 1 carbon is at the right.)
13. 3-butyl-4-ethyl-6-methyl-2-octene (The longest chain has 10 carbons, but the parent chain must contain the double bond.)
14. Polymerization is the process whereby a large molecule is assembled from many repeated smaller units.
15. butanol
16. styrofoam
17. a benzene ring (or a phenyl group)
18. C
19. $CH_3CH_2OCH_2CH_3$
20. CH_2O
21. $CH_3CH_2COCH_2CH_2CH_2CH_3$
22. $CH_3(CH_2)_6COOH$
23. $CH_3CH_2NH_2$
24. an amide
25. polyethylene, polystyrene, polypropylene, PVC, nylon, Teflon™, for example. "Plastic" will not suffice.
26. esters
27. LDPE, HDPE

Answers to Practice Test II

1. E
2. C_9H_{20}
3. branching
4. increase
5. Use the longest continuous chain of carbons (that has the double or triple bond if applicable).
6. alkynes
7. Dienes are hydrocarbons with two carbon-carbon double bonds.
8. $FCH_2CH{=}CHCH_3$
9.
```
CH3CH2CHCH2CH2CHCH2CH2CH3
      |        |
     H3C     CH3CCH3
               |
              CH3
```

10.
```
                CH3  CH3
                 |    |
CH3CH2CH2CHCHCH2CCH2CH2CH2CH3
             |        |
          CH3CH2     CH3
```

11.
```
           CH3CHCH3
              |
CH2=CHCH2CCH2CH2CH2CH3
              |
           CH3CHCH3
```

12.
```
CH3CH2        H
      \      /
       C = C
      /      \
     H        CH2CH3
```

13. 1,2-dibromopentane
14. amines
15. B (by elimination of the other answers, which are wrong!)
16. Aliphatic compounds have no benzene rings; aromatic compounds contain one or more benzene rings.
17. $CH_2OHCHOHCH_2OH$ or
```
CH2 CH CH2
 |   |   |
OH  OH  OH
```

18. G
19. B
20. D
21. F
22. C
23. A
24. E
25. A and B
26. C
27. D
28. A
29. B

CHAPTER 20

Biochemistry

Each of us is a chemical factory. All foods are chemicals. We are chemical. It is unfortunate that these terms come to have negative meaning. You know that a "chemical" substance is not a street drug; it is any substance.

You should want to learn the chemistry of your body, of your metabolism. Certainly we don't have time for too much detail, but what you learn in this chapter is a good base for future learning. There really is no "good" or "bad" food. Proper nutrition requires a *balance* of nutrients. If your nutrition is not balanced, your body will not produce necessary substances – proteins for growth and replacement of structure, vitamins, hormones, and many other compounds that are used for control of metabolism. You need all the *essential* vitamins, fatty acids, amino acids and minerals frequently. You also need carbohydrates, from which you obtain most of your energy and from which your body produces many of the substances it needs for metabolism. Nutritionists recommend diets high in *complex carbohydrate* (starches). These foods contain more than just the sugars that are the carbohydrates. Grains, for example, contain amino acids. Potatoes contain many minerals.

If you read the nutritional information on food labels, good! If you don't, start. Over time, you can learn a lot about nutrition by reading food labels, by reading the food sections of newspapers, and by listening to TV. You do have to use judgment. And you should realize that there simply is no magic bullet that will make you healthy.

SKILLS TO ENHANCE SUCCESS IN THIS CHAPTER

In this chapter the skill that you will need most is visualization of the microscopic world in three dimensions. The shapes of the molecules of compounds that are biologically active are just as important as their formulas. The direction of bonding is extremely important. Two sugars may have exactly the same molecular formula and be totally different to the body. You will learn of D- and L- designations for handedness. The molecules of D- and L-sugars are mirror images of one another. You will learn of α and β linkages, bonds differing only in the direction that the linkage projects out of the plane of the molecule.

A large number of new words and phrases are introduced. Biochemistry is complicated. It is organic chemistry plus. Try to learn the important factors regarding your metabolism, even if you don't remember each and every word used in this chapter. However, do try to learn the terminology sufficiently to recognize it when you read it or hear it.

CHAPTER 20 PROBLEMS

20.1 a The simplest sugars are *trioses*. Glyceraldehyde is important in the metabolism of carbohydrates. It is an *aldotriose*. Explain the term *aldotriose*. Draw glyceraldehyde.

Answer: The ending –ose means sugar. *Tri* means three. A *triose* is a three-carbon sugar, a sugar with three carbons in its molecule. *Aldo* means aldehyde. The structural formulas show the group –CHO on the end of the molecule. The aldehyde end of the sugar has a *carbonyl* group.

By convention, a linear structural formula of a monosaccharide is drawn vertically with the carbonyl carbon at or near the top of the structure.

CHO
H–C–OH
CH_2OH
Glyceraldehyde

20.1b. Use the aldohexose D-galactose as an example of a six-carbon sugar important to our metabolism. This, and glucose, from which we get our energy, are aldohexoses. Draw D-galactose. How are the carbons in sugars numbered? The bonding of which carbon determines the optical isomerism of the sugar? How is this indicated in the linear structural formula?

20.2a Write the condensed structural formula of the fatty acid called palmitic acid, $C_{15}H_{31}COOH$. Is it saturated or unsaturated? Is there an easy algebraic way to tell just by looking at the formula? How?
Answer: $CH_3CH_2CH_2CH_2CH_2CH_2CH_2CH_2CH_2CH_2CH_2CH_2CH_2CH_2CH_2COOH$
It is saturated. The number of hydrogens can be calculated from the length of the chain attached to the carboxylic acid. Ignoring the COOH part, there are 15 carbons in the chain. Each needs two hydrogens to be saturated, plus one more hydrogen for the end carbon. That means 2n + 1 hydrogens. $C_{15}H_{31}COOH$ has 15 carbons and 2n + 1 or 31 hydrogens, so it is saturated.

20.2b. Without drawing it, determine the number of double bonds, if any, in oleic acid, $C_{17}H_{33}COOH$. If it is not saturated, what would be the formula of the saturated fatty acid?

20.3 a The charges on the ends of an amino acid change with pH. With increasing pH, an amino acid will change in this way:

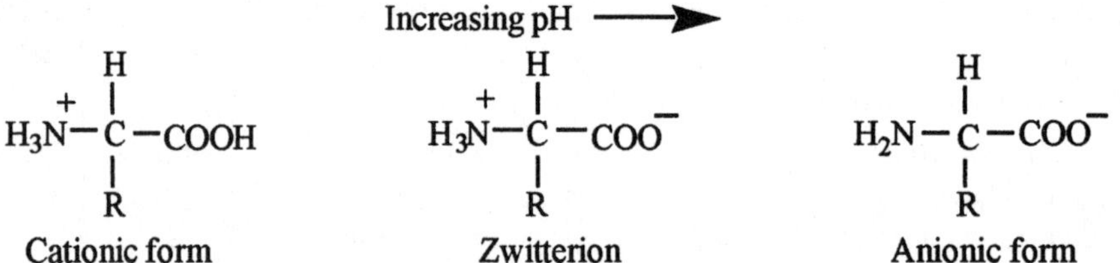

Enzymatic activity control by control of pH is an important control in human metabolism. Amino acids can also act as buffers. Show how this is possible.

20.3 b. Which amino acids would you expect to be acidic in pH? Use Table 20.3 of your text.

20.4 a Scientists around the world are busy trying to obtain the entire human *genome* – all the gene sequences that we use for our life processes. They use the abbreviations of the first letters in the base names as representing a peptide or protein.
Suppose that a sequence of mRNA was this:
-AUGCCGUCA–
What would be the sequence of the DNA produced from this mRNA?
Answer: –TACGGCAGT–

20.4 b. Draw the structure of pyrimidine. Which are the pyrimidine bases in nucleic acids?

20.5a. The nutritional informational on a granola bar is as follows:
Calories: 140 Fat: 3 g Carbohydrate: 27 g Protein: 2 g Calories from fat: 25
Each gram of carbohydrate and each gram of protein is worth 4 Cal. Each gram of fat is worth 9 Cal. Calculate the percentage of Calories from fat.
Answer:
Calories from fat = (3 g fat × 9 Cal/g) = 27 Cal from fat
Percentage Calories from fat = 27 Cal from fat / 140 total calories x 100% = 19% Calories from fat

20.5b. The protein requirements for an adult are 0.8 g protein per day per kg body weight. Calculate the requirement of protein each day for a 150-pound person. Assume that hamburger contains 25% water, 25% fat, and 50% protein on a pre-cooked basis. How much hamburger would supply one day 's protein?

Answers to "b" Problems

20.1b. The carbons are numbered from the end carbon closest to the carbonyl carbon. The carbonyl carbon of an aldose is carbon number 1.
The bonding of the next-to-last carbon determines whether the sugar is D or L. For a hexose, this carbon is carbon number 5. The hydroxyl group is drawn to the right on the formula, indicating a D configuration of the isomer.

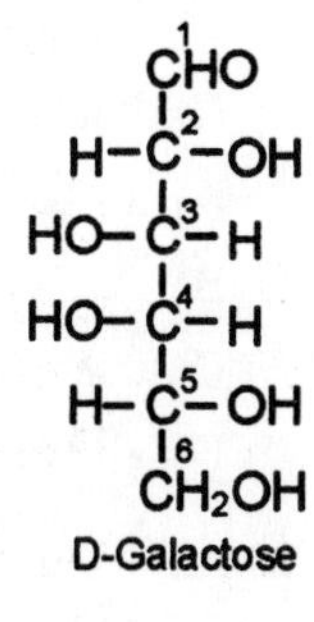

D-Galactose

20.2b. There is one double bond in oleic acid. We know this because the formula is two hydrogens short (at 33) of saturation (35). Every pair of missing hydrogens corresponds to one double bond between the carbons that are missing their hydrogens. The saturated fatty acid would be $C_{17}H_{35}COOH$. (#H's = 2n + 1)

20.3b. Amino acids with acid side chains (and acid names) would be acidic. Aspartic acid and glutamic acid have carboxylic acid groups.

20.4b. The structure of pyrimidine is shown below.

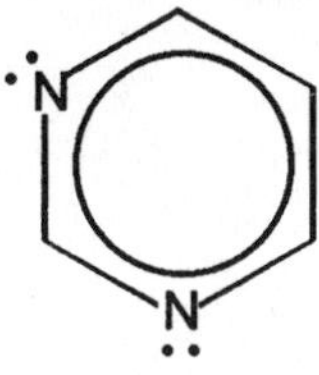

Pyrimidine

The three pyrimidine bases are cytosine, thymine, and uracil.

20.5b. The requirement is 54.5 grams of protein for a 150-pound (68.1 kg) person.
Assuming hamburger meat is 50% protein; 54.5 g of protein is provided by 109 grams of hamburger, or 0.24 pounds. A "quarter pounder" would do it!

PRACTICE TEST I

1. Monosaccharides can be classified as ____oses and _____oses.
2. The sugar $C_6H_{12}O_6$ is
   A) galactose.
   B) hexose.
   C) glucose.
   D) many different sugars.
   E) always a hemiacetal.
3. Polymeric glucose is
   A) protein. B) starch. C) lipid. D) cellulose. E) glycogen.
4. A 16-carbon saturated fatty acid has recently come under fire. This artery-hardening acid
   A) is found in palm oil.
   B) is linolenic acid.
   C) is found in beef tallow.
   D) is found in olive oil.
   E) is butyric acid.

5. A low iodine number means that the fat or oil is
   A) high in iodine content.
   B) more saturated.
   C) lower in calories.
   D) more likely to be solid.
   E) polyunsaturated.
6. The human body can synthesize
   A) all 20 amino acids (given energy).
   B) no amino acids; thus we must eat protein.
   C) over half of the amino acids.
   D) all but lysine.
7. DNA and RNA are composed of nucleotides, which in turn consist of which of the following?
   A) a heterocyclic base
   B) a nucleic acid
   C) a phosphate group
   D) a deoxy group
   E) a sugar
8. The secondary structure of the DNA is the famous ______ ______.
9. Which shows a proper RNA base-pairing sequence?

A)		B)		C)		D)	
	A – T		T – A		G – C		T – C
	G – C		A – T		C – G		A – G
	T – G		G – C		A – U		C – T
	G – A		C – G		C – G		A – G

10. Human insulin can now be produced in large quantities using
    A) nuclear reactors.
    B) cloned pancreases.
    C) chemical synthesis.
    D) recombinant DNA.
    E) algae.
11. The reason A T and G C pair up in DNA is
    A) to maximize hydrogen bonding between strands.
    B) that by bringing them close together, they react more quickly with each other.
    C) to connect the helices by double bonds.
    D) that they are of opposite charge.
12. A gene is
    A) composed of several chromosomes.
    B) a chunk of DNA.
    C) a base pair.
    D) a portion of a ribosome.
    E) a chunk of RNA.

13. Vitamins A, D, E, and K are __________–soluble vitamins.

14. Vitamin deficiencies can cause serious problems. Identify which vitamin deficiency causes each disease below.

Scurvy: ____________ deficiency

Pernicious anemia: ____________ deficiency

Rickets: ____________ deficiency

Beriberi: ____________ deficiency

Blindness: ____________ deficiency

In questions 15–19, match the physiological response to the hormone that regulates it.

15. Bone growth — A) insulin
16. Milk production — B) testosterone
17. Glucose usage — C) thyroxine
18. Metabolic rate — D) HGH
19. Aggressive behavior — E) prolactin

PRACTICE TEST II

1. Carbohydrates
   A) are hydrates of carbon.
   B) contain protein.
   C) are lipids.
   D) have the general formula $C_x(H_2O)_y$.
   E) are cyclic.
2. The process whereby the glucose molecule closes to two different ring forms and reopens is ________.
3. Our bodies cannot metabolize (digest and use)
   A) triglycerides.
   B) β polysaccharides.
   C) cyclic sugars.
   D) disaccharides.
   E) L-sugars.
4. Triglycerides with no double bonds are called
   A) vegetable oils.
   B) unsaturated fats.
   C) tridecenoic acids.
   D) saturated fats.
   E) alkanes.
5. A good test for "heart smart" oils can be done at home. Place oils in the refrigerator. The most __________ ones will solidify. These tend to cause arterial plaque.
6. Proteins are made up of ________________ linked by ___________ bonds.

7. DNA is made up of ______________ linked through the __________ group.

8. The famous double-helix structure was discovered by what pair of scientists?

9. A piece of one DNA strand consists of the nucleotides A-G-A-C-T. What will be the sequence of the complementary strand, written from left to right?

10. Write the name of the nucleotide base that corresponds to the letter abbreviations.

A ______________.

C ______________.

G ______________.

T ______________.

U ______________.

11. Which is toxic at high doses?

A) vitamin A

B) vitamin B_6

C) vitamin C

D) vitamin B_{12}

E) none of these

12. Hormones we require are

A) produced by leafy green vegetables.

B) all steroid derivatives.

C) found in meat and eggs.

D) synthesized in the liver.

E) produced by endocrine glands.

13. The hormone FSH (follicle-stimulating hormone) is produced in

A) the scalp.

B) the uterus.

C) all animal tissues.

D) the pituitary gland.

14. Testosterone is

A) normally present in large amounts.

B) produced in the bloodstream.

C) found in both males and females.

D) found in all animal and plant tissues.

15. Which is a steroid?

A) ibuprofen (an NSAID)

B) estradiol

C) vitamin C

D) insulin

E) vitamin E

16. Silk protein has the pleated sheet structure, which results in a stable, stretchy fiber. This protein has been sequenced and found to be rich in alanine ($R = CH_3$) and glycine ($R = H$), amino acids with small substituents on the α-carbon. How is the secondary structure, with its close stacking of sheets of protein chains, facilitated by the primary structure? How does the secondary structure result in the strength of a spiderweb or a parachute?

17. The base-pair codon of mRNA that calls for a methionine amino acid is "AUG". Transfer RNA, in order to bind the AUG codon, would have what base-pair sequence?

18. The bases A T G C are very hydrophobic. How is it then that DNA is water soluble?

19. According to Ralph Burns, author of your textbook, life without chemicals would be

A) anticlimactic.

B) boring.

C) cleaner (less polluted).

D) impossible.

E) easier to study for.

Answers to Practice Test I

1. aldoses and ketoses
2. D
3. B, D, and E
4. A
5. B
6. C
7. A, C, and E
8. double helix
9. C
10. D
11. A
12. B
13. fat
14. vitamin C
    vitamin B_{12}
    vitamin D
    thiamine (vitamin B_1)
    vitamin A
15. D (human growth hormone)
16. E
17. A
18. C
19. B

Answers to Practice Test II

1. D
2. mutarotation
3. B and E
4. D
5. saturated
6. amino acids, peptide (or perhaps amide)
7. nucleotides, phosphate
8. Watson and Crick
9. T-C-T-G-A
10. adenine
    cytosine
    guanine
    thymine
    uracil
11. A
12. E
13. D
14. C
15. B
16. The small R groups allow a close approach of the sheets. The close approach allows maximum hydrogen bonding between chains, thus a strong fiber.
17. UAC
18. The hydrophobic base pairs are inside the helix, with hydrophilic sugars and phosphates on the exterior.
19. D

CONGRATULATIONS! You're finished. Good luck in all your classes!

JR Frentrup *(Author of Practice Exams)*

Department of Chemistry

Eastern Michigan University

Ypsilanti, MI 48197

I welcome your written comments and suggestions.

TABLE OF CONTENTS

Chapter		
1	Chemistry is Everywhere	1
2	Matter and Energy	2
3	Fundamental Measurements	5
4	Elements, Atoms, and the Periodic Table	10
5	Atomic Structure: Atoms and Ions	15
6	Names, Formulas, and Users of Inorganic Compounds	19
7	Periodic Properties of Elements	22
8	Chemical Bonds	26
9	Chemical Quantities	30
10	Chemical Reactions	37
11	Stoichiometry: Calculations Based on Chemical Equations	42
12	Gases	48
13	Liquids and Solids	54
14	Solutions	58
15	Reactions Rates and Chemical Equilibrium	63
16	Acids and Bases	68
17	Oxidation and Reduction	74
18	Fundamentals of Nuclear Chemistry	82
19	Organic Chemistry	85
20	Biochemistry	90

CHAPTER 1
Chemistry is Everywhere

1.1 All items listed are made up of chemicals—they are composed of matter.

1.3 Benefit: Aspirin will reduce fever and pain. Risk: Aspirin may aggravate an ulcer condition.

1.5 There are many possible answers. Chemicals that save lives include medicines such as insulin and chemotherapy drugs used in treating disease. Chemicals that can be lethal include cocaine, heroin, and lead.

1.7 Chemistry deals with (a) the characteristics, (b) the composition, and (c) the chemical changes that occur in all material.

1.9 The metabolism and burning of carbohydrates are chemical changes.

1.11 Benefit: Burning fossil fuels provides energy for homes and industry.
Problem: Gaseous products of burning cause acid rain, an environmental problem. Chemists are studying these effects, reporting the problems, and suggesting solutions to save the environment.

1.13 Chemists study the chemistry of disease and develop chemicals (medicines) that are used in the diagnosis and treatment of disease: chemicals to combat infection, relieve pain, to control cancer, and to detect heart disease, AIDS, and so on.

1.15 Chemicals for high technology include materials for computer chips, digital displays, semiconductors, superconductors, audiotapes, laser discs, small movie cameras, fiber optics (used in communications applications), composite materials, and so on.

1.17 Chemists work in the production of all kinds of products and in the research and development of new products and analysis at all stages of the process, including environmental safety. Chemical engineers are involved in the design and day-to-day operation of chemical plants.

1.19 Five steps used in solving problems are summarized at the end of Sec. 1.2.

1.21 The law of gravity and the law of conservation of mass are natural laws; they summarize what happens repeatedly and consistently in nature.

1.23 Experiment, facts, hypothesis, theory, law.

1.25 Chemists in applied research are involved in the development of products—and solving problems related to the development of products—for business, industry, and the rest of society. Chemists in basic research are not looking for new products but, instead, are searching for solutions to unanswered questions. Much basic research is used by applied researchers.

1.27 a. basic research b. applied research

1.29 We must satisfy our own curiosity.

1.31 Experiments, Facts, Terminology, Laws, Theories, Problem Solving.

1.33 You will encounter chemistry in many other areas of science such as biology, physics and geology.

1.35 All involve chemistry.

CHAPTER 2
Matter and Energy

2.1 Earth's gravity is greater than the moon's, so the rock's weight is greater on Earth. The rock's mass is not affected by gravity. See example 2.1.

2.3 c. and d. contain matter; they have a mass and take up space, a. and b. do not.

2.5 a., b. and d. contain matter; they have a mass and take up space.

2.7 (a) miscible

2.9 liquid

2.11 (a) liquid, (b) liquid

2.13 See Table 2.3. Hydrogen and oxygen are colorless gases, but water is a liquid at room temperature.

2.15 See Table 2.3. Nitrogen and hydrogen are colorless, odorless gases at room temperature, but ammonia gas has a strong pungent and irritating odor.

2.17 An atom.

2.19 Elements, compounds, and water are substances. Light is not.

2.21 atom; mass

2.23 A compound.

2.25 Bronze is homogeneous. It has the same properties throughout.

2.27 Mouthwash is (d) a homogeneous mixture—a solution.

2.29 Table salt is (b) a compound.

2.31 Sulfur is an element.

2.33 Physical properties: a, b, c, e. Chemical properties: d.

2.35 Physical change: b, c. Chemical change: a, d.

2.37 Physical properties: a, d. Chemical properties: b, c.

2.39 Physical properties: color, odor, mass, length, waxy texture. Chemical properties: burns in the presence of oxygen.
Physical changes: wax melts. Chemical change: wax is burned to produce carbon dioxide and water.

2.41 Iron combines with oxygen in air to produce rust. The rust must have a mass equal to the sum of the mass of the iron plus the mass of the oxygen.

Thus, the mass of sulfur dioxide produced is greater than that of sulfur.

2.43 As gasoline burns, it combines with oxygen gas from the air. The sum of these masses equals the mass of the exhaust gases (carbon dioxide and water vapor) produced. The law of conservation of mass is not violated.

2.45 The roller coaster's potential energy is greatest at the top of the hill.

2.47 Exothermic: a, b, e Endothermic: c, d

2.49 b. released

2.51 Energy is either released or absorbed during a chemical change, but it is neither created nor destroyed.

2.53 When wood burns, chemicals in the wood combine with oxygen and energy is released—it is not created. The energy released is the difference between the chemical energy of the substances produced (the products) and the chemical energy of the substances that react (the wood and oxygen).

2.55 The solar energy absorbed by the plant is converted to potential energy.

2.57 $m = E/c^2$ For a small energy change, the change in mass is extremely small.

2.59 A chemical change occurs when wood burns. Energy is neither created nor destroyed. The amount of energy *released* is the difference between the large amount of energy in the reacting chemicals and the smaller amount of energy in the remaining substances.

2.61 c. Energy is neither created nor destroyed, but energy could be *released* or *absorbed* during a reaction.

2.63 Physical changes: a Chemical changes: b, c, d

2.65 Heterogeneous: b, d Homogeneous: a, c

2.67 Liquid water has more energy than ice. Molecules of the liquid are at a higher temperature and move more rapidly.

2.69 When stepping on the brake pedal, the kinetic energy of motion of the car is transferred into heat energy—as the brakes get hot—and into frictional energy that wears off the brake pads and tires. Energy is conserved as it is transformed from kinetic energy into heat and friction.

2.71 a, b

2.73 Energy is used to plant new trees, cut down the trees and process then into paper. Less energy is used and less materials are wasted in recycling paper products.

2.75 a. physical property; b. physical change; c. chemical change; d. chemical property

2.77 a. physical property; b. physical change; c. chemical change; d. chemical change

2.79 a. physical change; b. chemical change; c. physical change; d. chemical change

2.81 a. physical change; b. chemical change; c. physical change; d. physical change

2.83 a. decrease b. decrease c. increase

2.85 homogeneous; evaporation or distillation

2.87 Solid: definite shape and volume; liquid: indefinite shape and definite volume. Atomic level: In solids the particles are touching and are immobile. In liquids the particles are touching and mobile.

CHAPTER 3
Fundamental Measurements

3.1

numerical quantity		unit	substance
a.	1	gal	milk
b.	500	mg	vitamin C
c.	35	mm	film

3.3 a. 0.001, 0.001 b. 0.000001, 0.000001 c. 100, 100

3.5 a. (1) 1 mm b. (2) 3 cm c. (3) 22 cm

3.7 a. $0.062 \text{ m} \times \dfrac{100 \text{ cm}}{1 \text{ m}} = 6.2 \text{ cm}$

b. $3000. \text{ m} \times \dfrac{1 \text{ km}}{1000 \text{ m}} = 3.000 \text{ km}$

c. $875\ \mu\text{m} \times \dfrac{1 \text{ m}}{10^6\ \mu\text{m}} \times \dfrac{1 \text{ km}}{1000 \text{ m}} = 8.75 \times 10^{-7} \text{ km}$

3.9 a. $12.5 \text{ cm} \times \dfrac{1 \text{ m}}{100 \text{ cm}} \times \dfrac{1000 \text{ mm}}{1 \text{ m}} = 125 \text{ mm}$

b. $345 \text{ cm} \times \dfrac{1 \text{ m}}{100 \text{ cm}} = 3.45 \text{ m}$

c. $34.5 \text{ mm} \times \dfrac{1 \text{ m}}{1000 \text{ mm}} \times \dfrac{10^6\ \mu\text{m}}{1 \text{ m}} = 3.45 \times 10^4\ \mu\text{m}$

d. $10.5 \text{ mm} \times \dfrac{1 \text{ cm}}{10 \text{ mm}} = 1.05 \text{ cm}$

e. $42.5 \text{ m} \times \dfrac{100 \text{ cm}}{1 \text{ m}} = 4250 \text{ cm}$

f. $0.092 \text{ m} \times \dfrac{1000 \text{ mm}}{1 \text{ m}} = 92 \text{ mm}$

3.11 $670 \text{ nm} \times \dfrac{1 \text{ m}}{10^9 \text{ nm}} \times \dfrac{10^3 \text{ mm}}{1 \text{ m}} = 6.70 \times 10^{-4} \text{ mm}$

$670 \text{ nm} \times \dfrac{1 \text{ m}}{10^9 \text{ nm}} \times \dfrac{10^6\ \mu\text{m}}{1 \text{ m}} = 0.670\ \mu\text{m}$

3.13 0.475 μm; 4.75×10^{-4} mm

3.15 a. (2) 4 b. (3) 350 mL c. (2) 100. -mL cylinder

3.17 a. $0.050 \text{ L} \times \dfrac{1000 \text{ mL}}{1 \text{ L}} = 50. \text{ mL}$

b. $0.8\ \mu L \times \frac{1\ L}{10^6\ \mu L} \times \frac{1000\ mL}{1\ L} = 8 \times 10^{-4}\ mL$

c. $8.9\ cm^3 \times \frac{1\ mL}{1\ cm^3} \times = 8.9\ mL$

d. $75\ cc \times \frac{1\ L}{1000\ cc} = 0.075\ L$

3.19 $6.0\ m \times 0.50\ m \times 0.800\ m = 2.4\ m^3$

3.21 $2\ m^3 \times 10^3\ dm^3/m^3 = 2000\ dm^3$

3.23 a. \$1.79/2 L = \$0.895/L in bottles

b) $\frac{\$1.99}{6\ cans} \times \frac{1\ can}{0.345\ L} = \$0.937/L$ in cans

c. The cola in 2 L bottles is the better bargain.

3.25 a. (2) 3 cans of soda b. (3) 3 aspirins

3.27 a. $5.4\ g \times \frac{1000\ mg}{1\ g} = 5400\ mg$

b. $0.725\ kg \times \frac{1000\ g}{1\ kg} \times \frac{1000\ mg}{1\ g} = 7.25 \times 10^5\ mg$

c. $25\ \mu g \times \frac{1\ g}{10^6\ \mu g} = 2.5 \times 10^{-5}\ g$

d. $50.\ mL \times \frac{1\ g}{1\ mL} = 50.\ g$

3.29 0.045 g; $4.5 \times 10^4\ \mu g$

3.31 a. $165\ mm \times \frac{1\ cm}{10\ mm} \times \frac{1\ in.}{2.54\ cm} = 6.50\ in.$

b. $1200.\ mL \times \frac{1\ qt}{946\ mL} = 1.27\ qt$

c. $145\ lb \times \frac{1\ kg}{2.2\ lb} = 65.9\ kg$

d. $1.50\ ft \times \frac{12\ in.}{1\ ft} \times \frac{2.54\ cm}{1\ in.} = 45.7\ cm$

e. $500.\mu mL \times \frac{1\ fl\ oz}{29.6\ mL} = 16.9\ fl\ oz$

f. $275\ g \times \frac{1\ lb}{454\ g} = 0.606\ lb$

3.33 a. $6.5\ ft \times \frac{12\ in.}{1\ ft} \times \frac{1\ m}{39.4\ in.} = 2.0\ m$

b. $2.0 \text{ m} \times \frac{100 \text{ cm}}{1 \text{ m}} = 2\bar{0}0 \text{ cm}$

3.35 $\frac{1500. \text{ m}}{219 \text{ s}} = 6.85 \text{ m / s}$

3.37 $\frac{55 \text{ mi}}{3600 \text{ s}} \times \frac{1.61 \text{ km}}{1 \text{ mi}} \times \frac{1000 \text{ m}}{1 \text{ km}} = 24.6 \text{ m / s}$

3.39 $2.00 \times 10^5 \text{ s} \times \frac{1 \text{ min}}{60 \text{ s}} \times \frac{1 \text{ hr}}{60 \text{ min}} \times \frac{1 \text{ day}}{24 \text{ hr}} = 2.31 \text{ days}$

3.41 a. 3 b. uncertain c. 4 d. 6 e. 1 f. 3

3.43 a. 801 b. 0.0786 c. 0.0700 d. 7.10

3.45 a. 4.35×10^4 b. 6.50×10^{-4} c. 3.20×10^{-4} d. 4.32×10^2

3.47 a. 169.6 g b. 83 mm^3 c. 1.18×10^6

3.49 $2.0 \text{ cm} \times 3.5 \text{ cm} \times 0.52 \text{ cm} = 3.6 \text{ cm}^3$ $d = 1.53 \text{ g}/3.6 \text{ cm}^3 = 0.42 \text{ g/cm}^3$

3.51 $4000 \text{ cm}^3 \times \frac{11.4 \text{ g}}{1 \text{ cm}^3} \times \frac{1 \text{ kg}}{1000 \text{ g}} = 46 \text{ kg}$

3.53 $500. \text{ g} \times \frac{1 \text{ mL}}{0.79 \text{ g}} = 630 \text{ mL}$

3.55 a. $34.6 \text{ mL} - 24.0 \text{ mL} = 10.6 \text{ mL} = 10.6 \text{ cm}^3$ $d = 120.8 \text{ g}/10.6 \text{ cm}^3 = 11.4 \text{ g/cm}^3$

b. Lead

c. There are other metals with similar densities.

3.57 23.4478 g - 15.2132 g = 8.2346 g sample
25.9263 g - 15.2132 g = 10.7131 g water = 10.7131 mL water
8.2346 g/10.7131 mL = 0.76865 g/mL

3.59 $45 \text{ g} \times \frac{1 \text{ mL}}{1.19 \text{ g}} = 38 \text{ mL}$

3.61 $\frac{11.023 \text{ g sample}}{11.997 \text{ g water}} = 0.91881$ (specific gravity has no units)

3.63 $2.00 \text{ L} \times \frac{1.29 \text{ g}}{1 \text{ L}} = 2.58 \text{ g}$

3.65 a. $^{\circ}\text{C} = \frac{^{\circ}\text{F} - 32}{1.8} = \frac{25 - 32}{1.8} = -3.9^{\circ}\text{C}$ b. $^{\circ}\text{F} = (1.8 \times {}^{\circ}\text{C}) + 32 = (1.8 \times 20.) + 32 = 68\ {}^{\circ}\text{F}$

c. $°C = K - 273 = 298 - 273 = 25\ °C$

d. $°C = \dfrac{°F - 32}{1.8} = \dfrac{0. - 32}{1.8} = -18°C$

e. $°F = (1.8 \times °C) + 32 = (1.8 \times -40) + 32 = -40\ °F$

3.67 $°F = (1.8 \times °C) + 32 = [1.8 \times (-196)] + 32 = -321\ °F$

3.69 0 K, 0 °F, 0 °C

3.71 a. $1250\ \text{cal} \times \dfrac{1\ \text{Cal}}{1000\ \text{cal}} = 1.25\ \text{Cal}$

b. $1250\ \text{cal} \times \dfrac{4.18\ \text{J}}{1\ \text{cal}} \times \dfrac{1\ \text{kJ}}{1000\ \text{J}} = 5.23\ \text{kJ}$

3.73 $50.0\ \text{g} \times 30.0\ °C \times 1\ \text{cal/g}\cdot°C = 1500\ \text{cal}$

3.75 $100.\ \text{g} \times 70.0\ °C \times 0.448\ \text{J/g}\cdot°C = 3.14 \times 10^3\ \text{J}$

3.77 $\Delta T = 67.4\ °C - 21.2\ °C = 46.2\ °C$ $(4000\ \text{g})(1.00\ \text{cal/g}\cdot°C)(46.2\ °C) = 185000\ \text{cal} = 185\ \text{Cal}$

3.79 a. $5.00\ \text{gal} \times \dfrac{4\ \text{qt}}{1\ \text{gal}} \times \dfrac{946\ \text{mL}}{1\ \text{qt}} \times \dfrac{13.55\ \text{g}}{1\ \text{mL}} \times \dfrac{1\ \text{kg}}{1000\ \text{g}} = 256\ \text{kg}$

b. $256\ \text{kg} \times \dfrac{2.20\ \text{lb}}{1\ \text{kg}} = 563\ \text{lb}$

3.81 a. $\dfrac{200.\ \text{mg}}{1\ \text{day}} \times \dfrac{1\ \text{glass}}{62.0\ \text{mg}} = 3.23\ \text{glass / day}$

b. $4\ \text{people} \times 7\ \text{days} \times \dfrac{3.23\ \text{glasses}}{1\ \text{day}} \times \dfrac{4\ \text{oz}}{1\ \text{glass}} \times \dfrac{1\ \text{qt}}{32\ \text{oz}} = 11.3\ \text{qt}$

3.83 a. $4\ \text{slices} \times \dfrac{495\ \text{Cal}}{1\ \text{slice}} = 1980\ \text{Cal}$ $\dfrac{1980\ \text{Cal}}{2250\ \text{Cal}} \times 100\% = 88.0\ \%$

b. $\dfrac{495\ \text{Cal}}{2250\ \text{Cal}} \times 100\% = 22.0\ \%$ $0.220 \times 24\ \text{hrs} = 5.28\ \text{hr}$

3.85 a. $202000\ \text{cal} \times 4.18\ \text{J/cal} = 8.44 \times 10^5\ \text{J}$

b. $\dfrac{8.44 \times 10^5 \text{J}}{(17°C)(4.18\ \text{J/g}°C)} = 1.19 \times 10^4\ \text{g}$

3.87 $(3.80 \times 10^3\ \text{cm}^3) \times 1\ \text{in.}^3/16.4\ \text{cm}^3 = 232\ \text{in.}^3$

3.89 a. $(11.1\ \text{in.})^3 \times \dfrac{16.4\ \text{cm}^3}{1\ \text{in.}^3} \times \dfrac{1\ \text{L}}{1000\ \text{cm}^3} = 22.4\ \text{L}$

b. $22.4\ \text{L} \times \dfrac{1.168\ \text{g}}{1\ \text{L}} = 26.2\ \text{g}$

3.91 a. $1\ \text{week} \times \dfrac{7\ \text{day}}{1\ \text{wk}} \times \dfrac{24\ \text{hr}}{1\ \text{day}} \times \dfrac{60\ \text{min}}{1\ \text{hr}} \times \dfrac{60\ \text{s}}{1\ \text{min}} \times \dfrac{1\ \text{drop}}{1\ \text{s}} \times \dfrac{1\ \text{mL}}{20\ \text{drops}} \times \dfrac{1\ \text{L}}{1000\ \text{mL}} = 15\ \text{L}$

b. $15\ \text{L} \times \dfrac{1\ \text{qt}}{0.946\ \text{L}} \times \dfrac{1\ \text{gal}}{4\ \text{qt}} = 4.0\ \text{gal}$

3.93 The term melting refers to the change of a substance from a solid to a liquid. This is different from dissolving a solid in a liquid.

3.95 $\frac{1\,\text{molecule}}{3.0\text{nm}} \times \frac{10^9\,\text{nm}}{1\text{m}} \times \frac{1\text{m}}{100\text{cm}} \times \frac{2.54\text{cm}}{1\text{inch}} = 8.5\text{x}10^6\,\text{molecules/inch}$

3.97 a. $10.0\,\text{km} \times \frac{0.6215\,\text{mile}}{1\,\text{km}} = 6.22\,\text{mile}$

b. $16.0\,\text{fl oz} \times \frac{29.6\,\text{mL}}{1\,\text{fl oz}} = 474\,\text{mL}$

c. $16.0\,\text{oz} \times \frac{28.35\,\text{g}}{1\,\text{oz}} = 454\,\text{g}$

3.99 $250.\,\text{tablets} \times \frac{30.0\,\text{mg}}{1\,\text{tablet}} = 75\bar{0}0\,\text{mg}$ or $7.50\,\text{g}$

3.101 $50.\,\text{nm} \times \frac{1\,\text{m}}{10^9\,\text{nm}} \times \frac{100\,\text{cm}}{1\,\text{m}} = 5.0 \times 10^{-6}\,\text{cm}$

3.103 $\frac{435\,\text{mg}}{1\,\text{dL}} \times \frac{1\,\text{g}}{1000\,\text{mg}} \times \frac{10\,\text{dL}}{1\,\text{L}} = 4.35\,\text{g/L}$

CHAPTER 4
Elements and Atoms

4.1 Alchemists were the experimentalists of the Middle Ages; they attempted—among other things—to achieve the transmutation of various metals into gold.

4.3 According to Boyle, an element is a substance that cannot be broken down to give simpler substances.

4.5 Antoine Lavoisier

4.7 e. technetium

4.9 The first letter of a symbol for an element is always capitalized.

4.11 a. K b. Mn c. Cu d. Au e. P f. F

4.13 a. arsenic b. barium c. antimony d. silicon e. platinum f. nitrogen

4.15 See Table 4.1

4.17 b. hydrogen

4.19 Hydrogen 93%, helium 7%, 0.1% others. Hydrogen and helium are the first two elements on the periodic table.

4.21 a. metalloid b. metal c. nonmetal d. metal e. nonmetal f. nonmetal

4.23 Non-metals are usually dull appearance and typically softer than metals. Nonmetals are not malleable.

4.25 c, d

4.27 Hydrogen is a gas at room temperature. It is colorless, odorless and flammable.

4.29 Hydrogen, nitrogen, oxygen, fluorine, and chlorine. All are nonmetals.

4.31 b and e, sulfur and phosphorus

4.33 b. phosphorus

4.35 Leucippus and Democritus held the atomistic view of matter. Aristotle and most ancient Greek philosophers believed matter was continuous.

4.37 J. Priestley and W. Scheele discovered oxygen, independently. Priestley, whose discovery was announced first, is usually credited with the discovery.

4.39 Lavoisier

4.41 The law of definite proportions is supported by the data. The hydrogen to oxygen volume ratio is 2:1 in both cases.

4.43 In Berzelius' experiment, a given mass of lead always combined with a specific mass of sulfur. This can be explained by Dalton's theory that a given number of atoms of lead must always combine with a fixed number of atoms of sulfur. Using an excess of sulfur could not give more lead sulfide.

4.45 a. The law of definite proportions.

b. The law of multiple proportions and the law of definite proportions.

c. The Cl mass ratio is 2:1 for Sample B to Sample A. Cu is fixed.

4.47 $10.5 \text{ g N} \times \frac{32.0 \text{ g O}}{14.0 \text{ g N}} = 24.0 \text{ g O}$

4.49 $6 \times 10^{10} \text{ C atoms} \times \frac{4 \text{ H atoms}}{1 \text{ C atom}} = 2.4 \times 10^{11} \text{ H atoms}$

This is in accordance with the law of definite proportions.

4.51 A scientific law merely summarizes experimental fact, often in mathematical fashion. Governmental laws merely are rules that attempt to control behavior and dictate what should be; they do not summarize scientific facts.

4.53 1) "All elements are made of tiny, indivisible particles called atoms. Atoms can be neither created or destroyed during chemical reactions." The same particles (atoms) are present before and after a reaction.

4.55 Proust's Law of definite proportions. Atoms combine in simple whole number ratios.

4.57 4) " If the same elements form more than one compound, there is a different but definite, small whole number mass ratio and atom ratio for each compound." The ratios are in whole numbers.

4.59 A proton has a positive charge and a mass number of 1 amu. An electron has a negative charge and a mass number of 0 amu. A neutron has no charge but has a mass number of 1 amu.

4.61 Atoms are neutral because they contain equal numbers of protons (positive) and electrons (negative). The protons are located in an atom's nucleus. The electrons are located in the region outside an atom's nucleus.

4.63 Elements, protons, electrons (in that order): a. Ca, 20, 20 b. Pb, 82, 82 c. P, 15, 15 d. Ne, 10, 10

4.65 a. 0 b. 17, 17, 18

4.67 a. 0 b. 35, 35, 46

4.69 a. no b. yes c. 21 amu d. 20 amu

4.71 ${}^{2}_{1}\text{H}$, atomic number =1, mass number = 2, 1 proton, 1 neutron, and 1 electron

4.73 a. 86 b. 136 c. 86

4.75 a. 26 b. 33 c. 26

4.77
$$\begin{aligned} 10.0129 \text{ amu} \times 0.200 &= 2.00 \text{ amu} \\ 11.0093 \text{ amu} \times 0.800 &= \underline{8.81 \text{ amu}} \\ & 10.81 \text{ amu} \end{aligned}$$

4.79 23.9850 amu x 0.7899 = 18.95 amu
24.9850 amu x 0.1000 = 2.499 amu
25.9826 amu x 0.1101 = 2.861 amu
24.31 amu

4.81 a. 226.13 g/100 dimes = 2.26 g/dime

b. 12.00 g / 6.02 x 10^{23} C atoms = 1.99 x 10^{-23} g/C atom

4.83 a. 40.08 g/mol b. 6.02 x 10^{23} atoms

4.85 a. 26.98 g/mol b. 6.02 x 10^{23} atoms

4.87 a. Mg 1 x 24.3 g = 24.3 g
O 2 x 16.0 g = 32.0 g
H 2 x 1.0 g = 2.0 g
58.3 g/mol

b. $1.27 \text{ mol} \times \frac{58.3 \text{ g } Mg(OH)_2}{1 \text{ mol}} = 74.0 \text{ g } Mg(OH)_2$

4.89 a. Ca 1 x 40.1 g = 40.1 g
C 1 x 12.0 g = 12.0 g
O 3 x 16.0 g = 48.0 g
100.1 g/mol

b. N 2 x 14.0 g = 28.0 g
H 4 x 1.0 g = 4.0 g
O 3 x 16.0 g = 48.0 g
80.0 g/mol

c. Na 3 x 23.0 g = 69.0 g
P 1 x 31.0 g = 31.0 g
O 4 x 16.0 g = 64.0 g
164.0 g/mol

4.91 a. 1.22 mole $CaCO_3$ × 100.1 g/mole = 122 g $CaCO_3$

b. 1.22 mole NH_4NO_3 × 80.0 g/mole = 97.6 g NH_4NO_3

c. 1.22 mole Na_3PO_4 × 164.0 g/mole = 200. g Na_3PO_4

4.93 a. 10.00 g Pb x 1 mol Pb/ 207.2 g Pb = 0.0483 mol Pb
1.56 g S x 1 mol S/32.1 g S = 0.0486 mol S

b. 1:1 mole ratio

c. points 3) and 5); Atoms of the elements combine in small whole number fixed ratios to form compounds.

4.95 a. $53 \text{ g Cu} \times \frac{1 \text{ mol Cu}}{63.6 \text{ g Cu}} = 0.833 \text{ mol Cu}$

b. $40 \text{ g O} \times \frac{1 \text{ mol O}}{16.0 \text{ g O}} = 2.50 \text{ mol O}$

c. $10 \text{ g C} \times \frac{1 \text{ mol C}}{12.0 \text{ g C}} = 0.833 \text{ mol C}$

d. 1:1:3 $CuCO_3$

4.97 a. Co is the symbol for the element cobalt; CO is the formula for the compound carbon monoxide.

b. Pb is the symbol for the element lead; PB is the formula for a compound of phosphorus and of boron.

c. $Ca(OH)_2$ is the formula for a compound of calcium, oxygen and hydrogen.

4.99 During combustion, the total mass (counting all gases, liquids and solids) remains unchanged, and obeys the law of conservation of mass. The apparent loss of mass is because the masses given do not include masses of all substances (gases) present before and after the wood is burned.

4.101 $10.0 \text{ g } H_2 \times \frac{36.0 \text{ g water}}{4.0 \text{ g } H_2} = 90.0 \text{ g water}$

4.103 a. $0.750 \text{ mol Hg} \times \frac{200.6 \text{ Hg}}{1 \text{ mol Hg}} = 150. \text{ g Hg}$

$0.750 \text{ mol Hg} \times \frac{6.02 \times 10^{23} \text{ Hg atoms}}{1 \text{ mol Hg}} = 4.52 \times 10^{23} \text{ Hg atoms}$

b. $0.750 \text{ mol Cr} \times \frac{52.0 \text{ g Cr}}{1 \text{ mol Cr}} = 39.0 \text{ g Cr}$

$0.750 \text{ mol Cr} \times \frac{6.02 \times 10^{23} \text{ Cr atoms}}{1 \text{ mol Cr}} = 4.52 \times 10^{23} \text{ Cr atoms}$

4.105 $1.00 \text{ mm} \times \frac{1 \text{ m}}{1000 \text{ mm}} \times \frac{1 \text{ atom}}{10^{-10} \text{ m}} = 10^7 \text{ atoms}$ (10 million atoms)

4.107 $^{99}_{43}Tc$ has 43 protons, 56 neutrons, and 43 electrons. The atomic number is 43 and the mass number is 99.

4.109 a. Cr-52 p = 24, e=24, n= 28

b. Ar-40 p=18, e=18, n=22

4.111 1 mol of sulfur trioxide, SO_3, has 6.02×10^{23} molecules.

4.113 a. 12 + 22 + 11 = 45 atoms/formula unit

b. $\frac{45 \text{ atoms}}{\text{formula unit}} \times \frac{6.02 \times 10^{23} \text{ formula units}}{\text{mol}} = 2.7 \times 10^{25} \text{ atoms/mol}$

c.

C	12	x 12.0 g	=	144.0 g
H	22	x 1.0 g	=	22.0 g
O	11	x 16.0 g	=	176.0 g
				342.0 g/mol

4.115 Atomic number = 26; protons = 26; electrons = 26; neutrons = 33; mass number = 59; electrical charge = 0

4.117 a. Fe and Br combined in more than one set of proportions.

CHAPTER 5
Atomic Structure: Atoms and Ions

5.1 Mendeleev published a chemistry textbook with a periodic table that showed elements arranged in order of increasing atomic masses which, in turn, had been determined rather accurately by Dulong, Petit and others.

5.3 A beam of light, called a cathode ray, passed between the cathode and the anode in a straight line.

5.5 A modern fluorescent tube is closely related to the Crookes tube.

5.7 The *e/m* ratio (the charge to mass ratio) for an electron was first determined by J. J. Thomson. Knowing this ratio was one step closer to knowing the mass of the electron which could be calculated once the charge, *e*, of the electron was determined.

5.9 In Thomson's "plum pudding" model of the atom, the neutral atom has an equal number of negative charges (electrons) and positive charges (protons) spread throughout the atom. Later, Rutherford's experiments showed that the positive charge is not spread out; it is concentrated in the nucleus of each atom.

5.11 Millikan was able to determine the charge of an electron.

5.13 X-rays

5.15 See Figure 5.9 and Table 5.1.

5.17

Observation	Conclusion
(1) Most of the alpha particles went through the gold foil.	The mass is concentrated into a tiny nucleus; the atom is mostly empty space.
(2) Some of the alpha particles were deflected.	Alpha particles passing near the nucleus are deflected.
(3) A few alpha particles bounced back.	Alpha particles bounce backwards when they hit the nucleus directly.

Model: The positive charge and nearly all the mass of an atom are contained in the tiny nucleus of the atom.

5.19 Lower frequencies than visible light: infrared, radar, TV/FM, short wave radio.

5.21 Visible light (highest frequency first): violet, indigo, blue, green, yellow, orange and red.

5.23 Ultraviolet light, UV, is often called "black light"

5.25 radar

5.27 High to low frequencies: UV, visible light, shortwave radio.

5.29 infrared, IR.

5.31 short wavelength in the TV/FM region

5.33 Each element has a characteristic line spectrum; that is, a specific set of frequencies that are emitted by atoms of an excited (energized) element.

5.35 Sodium. Sodium compounds are present in wood and all other plant cells.

5.37 $E = h\nu$. Energy and frequency increase proportionally.

5.39 Barium or copper

5.41 An atom in its "ground state" has all of its electrons in their lowest energy levels. In an "excited state" one or more electrons is elevated to a higher energy level.

5.43 heat, light, electron bombardment, and chemical reactions

5.45 Light is emitted when excited electrons within an atom fall back to lower energy states.

5.47 The first, second, third, and fourth energy levels can have 2, 8, 18, and 32 electrons, respectively.

5.49 Bohr diagrams show concentric rings to represent energy levels of an atom or ion, along with the appropriate number of electrons in each energy level. Listed here are the symbols of the elements with the number of electrons in each energy level, beginning at the nucleus. The valence electrons are those in the outermost energy level.

a. Mg: 2, 8, 2	Each atom has 2 valence electrons.
b. Ca: 2, 8, 8, 2	Each atom has 2 valence electrons.
c. N: 2, 5	Each atom has 5 valence electrons.
d. S: 2, 8, 6	Each atom has 6 valence electrons.
e. F: 2, 7	Each atom has 7 valence electrons.

5.51

a.	K^+	$2e^-$	$8e^-$	$8e^-$	A potassium atom lost one electron.
b.	Cl^-	$2e^-$	$8e^-$	$8e^-$	A chlorine atom gained one electron.

5.53 Electrons can only be in specific energy levels and not in between.

5.55 The atom must: a. gain one e^- b. lose $2e^-$ c. lose 3 e^-

5.57

Level	a. Si	b. P	c. Al	d. Ar	e. K
1	2	2	2	2	2
2	8	8	8	8	8
3	4	5	3	8	8
4					1

5.59 Schrödinger and Heisenberg showed, mathematically, that electrons do not exist in simple, spherical planetary orbits. Thus, we need to think of electrons in terms of electron clouds that occupy general regions—with various shapes—rather than of electrons in simple spherical orbits.

5.61 a. K• b. •B• c. •N• d. •S• e. :Cl•

5.63 a. :F:$^-$ b. Ca^{2+} c. :S:$^{2-}$

5.65 In the second energy level there are two sublevels, *s* and *p*. The *s* sublevel has one *s* orbital; the *p* sublevel has three *p* orbitals.

5.67 A *d* sublevel is partially filled for transition elements. The energy level number of the *d* sublevel is one lower than the energy level for *s* and *p* block elements of the same period. In filling sublevels, 3*d* follows 4*s*; 4*d* follows 5*s*, etc.

5.69

		2s	2p		
a. Li $1s^2 2s^1$	[He]	↑	__	__	__
		3s	3p		
b. Al $1s^2 2s^2 2p^6 3s^2 3p^1$	[Ne]	↑↓	↑	__	__
c. P $1s^2 2s^2 2p^6 3s^2 3p^3$	[Ne]	↑↓	↑	↑	↑
		2s	2p		
d. O $1s^2 2s^2 2p^4$	[He]	↑↓	↑↓	↑	↑

		4s	3d					4p		
e. Br [Ar]$4s^2 3d^{10} 4p^5$	[Ar]	↑↓	↑↓	↑↓	↑↓	↑↓	↑↓	↑↓	↑↓	↑

5.71

		2s	2p		
a. F^- $1s^2 2s^2 2p^6$	[He]	↑↓	↑↓	↑↓	↑↓
		3s	3p		
b. Ca^{2+} $1s^2 2s^2 2p^6 3s^2 3p^6$	[Ne]	↑↓	↑↓	↑↓	↑↓
c. S^{2-} $1s^2 2s^2 2p^6 3s^2 3p^6$	[Ne]	↑↓	↑↓	↑↓	↑↓

5.73 de Broglie stated that a beam of electrons should exhibit wave characteristics. This idea was supported by calculations by Schrodinger which combine the wave properties and the particle nature of an electron. Heisenberg also supported the wave theory with his uncertainty principle which states the specific path of an electron cannot be determined. If the electron acts like a particle it's position should be able to be determined.

5.75 In the Rutherford experiment, alpha particles were used to bombard gold foil. These alpha rays were produced by samples of naturally radioactive Radium which had been discovered and isolated by the Curies.

5.77 $\lambda = c/\nu$ $\quad \lambda = (3.00 \times 10^8 \text{ m/s})/(900 \times 10^6 \text{s}^{-1}) = 0.333 \text{ m}$ or 33.3 cm

5.79 a. Dalton did not know about any of the subatomic particles or isotopes.

b. Thomson did not know about the nucleus of an atom, where all of the positive charge and nearly all of the mass of an atom is concentrated.

c. Bohr knew about energy levels of electrons, but he did not know about sublevels and orbitals and their shapes.

5.81 Each hydrogen atom can have only one electron in an excited state but for any sample there is not just one excited atom, or one excited state. The excited electrons in different atoms can fall back to several lower states before reaching the ground state. Each electron transition corresponds to a specific frequency and a different line in the spectrum.

5.83 These elements all have 1 valence electron; they are called the alkali metals.

5.85 2 (the electrons in one of the 2*p* orbitals are paired; the other two are unpaired)

5.87 They each have two valence electrons. They are all in the second group in the table.

5.89 a. aluminum - it has 3 valence electrons (in the 3*s* and 3*p* sublevels)

b. fluorine - it has 7 valence electrons (in the 2*s* and 2*p* sublevels)

			4*s*	3*d*				
5.91	a. Co [Ar]$4s^2 3d^7$	[Ar]	↑↓	↑↓	↑↓	↑	↑	↑
	b. Ni [Ar]$4s^2 3d^8$	[Ar]	↑↓	↑↓	↑↓	↑↓	↑	↑
	c. Zn [Ar]$4s^2 3d^{10}$	[Ar]	↑↓	↑↓	↑↓	↑↓	↑↓	↑↓

5.93 They all have 2 valence electrons. They are the alkaline earth metals.

5.95 Heat

5.97 UV-A

5.99 3*d* and 4*d*

5.101 4*f* and 5*f*

5.103 Sr atom: $1s^2 2s^2 2p^6 3s^2 3p^6 4s^2 3d^{10} 4p^6 5s^2$ [Kr] ↑↓

Sr^{2+} ion: $1s^2 2s^2 2p^6 3s^2 3p^6 4s^2 3d^{10} 4p^6$ [Kr]

CHAPTER 6
Names, Formulas, and Uses of Inorganic Compounds

6.1 oxygen

6.3 a. NH_4^+, 5 atoms b. HSO_4^-, 6 atoms c. HSO_3^-, 5 atoms

d. MnO_4^-, 5 atoms e. OH^-, 2 atoms f. $Cr_2O_7^{2-}$, 9 atoms

g. ClO^-, 2 atoms h. ClO_4^-, 5 atoms

6.5 a. Cu^+, cuprous and Cu^{2+}, cupric. The - *ic* ending designates the ion with the higher charge number.

b. Fe^{2+}, ferrous and Fe^{3+}, ferric. The *-ic* ending designates the ion with the higher charge number.

6.7 O_2^{2-} ; Hg_2^{2+}

6.9 a. chlorate b. perchlorate c. carbonate d. thiocyanate e. sulfate.

6.11 a. Na^+, O_2^{2-} , Na_2O_2 b. Fe^{3+} , O^{2-} , Fe_2O_3

c. Co^{3+} , NO_3^- , $Co(NO_3)_3$ d. Cr^{3+} , SO_4^{2-} , $Cr_2(SO_4)_3$

e. K^+ , MnO_4^- , $KMnO_4$

6.13 a. Ca^{2+}, OH^-, $Ca(OH)_2$ b. Sn^{2+}, F^- , SnF_2

c. NH_4^+, PO_4^{3-} , $(NH_4)_3PO_4$ d. Ca^{2+}, CO_3^{2-} , $CaCO_3$

e. Ca^{2+}, ClO^- , $Ca(ClO)_2$

6.15 a. tin(IV) chloride b. mercury(I) chloride c. iron(II) oxide d. manganese(II) chloride
e. copper(II) sulfide f. cobalt(III) nitrate

6.17 a. stannic chloride b. mercurous chloride c. ferrous oxide d. manganous chloride
e. cupric sulfide f. cobaltic nitrate

6.19 a. $Ba(OH)_2$ b. $Cr(NO_2)_3$ c. $(NH_4)_2CO_3$ d. $Cu(CN)_2$ e. $Ca_3(PO_4)_2$

6.21 a. $Sn(CrO_4)_2$ b. PbS c. $Fe(CH_3COO)_2$ d. NH_4SCN e. $Ni(OH)_3$

6.23 a. NO b. CS_2 c. dinitrogen trioxide d. dinitrogen monoxide e. sulfur trioxide
f. diphosphorus pentoxide

6.25 a. PCl_5 b. N_2O_4 c. tetraphosphorus decoxide d. dinitrogen pentoxide

6.27 a. $2(+1) + S + 4(-2) = 0 \quad S = +6$

b. $S + 3(-2) = -2 \quad S = +4$

c. $S + 3(-2) = 0$ $S = +6$

d. $+1 + I + 3(-2) = 0$ $I = +5$

e. $+1 + I + 2(-2) = 0$ $I = +3$

f. $+1 + I + (-2) = 0$ $I = +1$

6.29 a. HBr b. HNO_3 c. HNO_2 d. H_2CO_3 e. AgBr f. $AgNO_3$

6.31 a. CH_3COOH b. KCH_3COO c. H_2SO_3 d. $(NH_4)_2SO_3$ e. HCl(aq) f. $MgCl_2$

6.33 a. phosphoric acid b. potassium phosphate c. potassium hydrogen phosphate
d. potassium dihydrogen phosphate

6.35 a. carbonic acid b. potassium bicarbonate c. potassium carbonate d. hydrofluoric acid

6.37 a. $Na_2CO_3 \bullet 10H_2O$, sodium carbonate decahydrate, water softening

b. $Na_2S_2O_3 \bullet 5H_2O$. sodium thiosulfate pentahydrate, photographic developer

c. $Ca(OH)_2$, calcium hydroxide, mortar and neutralizing acids

d. $Mg(OH)_2$, magnesium hydroxide, antacid and laxative

e. $NaHCO_3$, sodium bicarbonate, baking powder, cooking

f. H_2SO_4, sulfuric acid, manufacture of fertilizers and chemicals

g. CO_2, carbon dioxide, fire extinguishers and refrigeration

h. N_2O, dinitrogen monoxide (nitrous oxide), anesthesia and fuel

6.39 a. H_2SO_4, oxidation of sulfur, SO_3 + water, 48 million tons, manufacture of fertilizers and chemicals

b. H_3PO_4, phosphate rock + acid, 13 million tons, fertilizers and detergents

c. NaOH, electrolysis of NaCl, 12 million tons, chemical and paper manufacturing

d. Na_2CO_3, minerals or brine, 11 million tons, glass and chemicals

6.41 $20 \times 10^6 \text{ tons} \times \frac{\$135}{1 \text{ ton}} = \$2.7 \times 10^9$ about 2.7 billion dollars

6.43 The cation.

6.45 All acids contain ionizable hydrogens. In water solution, each ionizable hydrogen becomes a hydronium ion, H_3O^+.

6.47 CO is carbon monoxide. Co is the symbol for cobalt.

6.49 Ammonia gas is NH_3; the ammonium ion, NH_4^+, has a proton, H^+, added.

6.51 SO_3 is the formula for sulfur trioxide and SO_3^{2-} is the formula for the sulfite ion.

6.53 Parenthesis are placed around a polyatomic ion in a formula when—and only when—two or more of the polyatomic ions are represented in the formula.

6.55 a. RbI b. Ra_3N_2 c. CoP d. CaSe

6.57 A hydrate is a crystalline compound that contains a definite, fixed number of water molecules.

6.59 a. 4 b. 3 c. 2

6.61 sodium chloride, NaCl; magnesium chloride, $MgCl_2$; aluminum chloride, $AlCl_3$; silicon tetrachloride, $SiCl_4$; phosphorus trichloride, PCl_3; sulfur dichloride, SCl_2; chlorine, Cl_2.

6.63 a. HCN(g) b. HI(aq) c. Na_2O_2 d. $NaKSO_4$ e. $Na_2S_2O_3 \cdot 5H_2O$

6.65 a. H_2S b. Ag_2S c. $MgCl_2 \cdot 6H_2O$ d. $CaCO_3$ e. WC

6.67 a. sodium thiocyanate b. nitrogen dioxide c. nitrite ion

6.69 a. ammonium oxalate b. magnesium bicarbonate c. phosphorous acid

6.71 a. binary covalent compound b. acid c. salt

6.73 a. carbon dioxide b. cobalt(II) oxide c. carbon monoxide

6.75 a. +6 b. +5 c. +4

6.77 a. barium peroxide b. carbon dioxide c. carbonic acid
d. ammonium hydrogen carbonate

6.79 nitrogen trifluoride

6.81 nitrate

CHAPTER 7
Periodic Properties of the Elements

7.1 When atomic masses for Ca and Ba are averaged, the result is very close to the atomic mass of strontium (87.6).

7.3 Mendeleev is generally credited with discovery of the periodic table. Although the periodic table developed by Meyer was submitted for publication before Mendeleev's, it was not published until after Mendeleev's table was published. Furthermore, Mendeleev gained notoriety by predicting properties of certain undiscovered elements.

7.5 Mendeleev left gaps in his periodic table for unknown elements. He boldly predicted that elements would be discovered with various predicted properties.

7.7 Atomic mass

7.9 Atomic numbers were not known when Mendeleev developed his periodic table. When elements are arranged by atomic number, Te, I, Co, Ni, Ar, and K fall in appropriate families in the periodic table; this was not the case when atomic mass was used.

7.11 The terms "family of elements" and "groups of elements" both refer to the vertical columns of elements in the periodic table. When "group" is used, a number is given such as Group IA or Group VIIA. When "family" is used, a family name is given such as the halogen family (Group VIIA).

7.13 For the second and third periods of elements, the shiny reactive metals are at the left, followed by dull solids, reactive nonmetals, and a non-reactive noble gas. This accompanies the metallic to nonmetallic trend.

7.15 Each element in Group VIA (the oxygen family) has 6 valence electrons.

7.17 a. F and Br are in Group VIIA and have seven valence electrons.

b. Mg and Ca are in Group IIA and have two valence electrons.

c. C and Si are in Group IVA and have four valence electrons.

d. He and Ar are in Group VIIIA and have two and eight valence electrons respectively.

7.19 The alkali metals are in Group IA, the alkaline earth metals are in Group IIA, and the halogens are in Group VIIA.

7.21 Period 1 has two elements. Period 2 has eight elements.

7.23 Group IA elements have the greatest atomic size (Figure 7.7).

7.25 Atomic size within a period decreases as atomic number increases. This is because each element within a period has one more proton than the previous element and the increase in positive charge draws the electron cloud closer to the nucleus.

7.27 a. K atom p=19, n=20, e=19; K^+ ion p=19, n=20, e=18

b. Br atom p=35, n=45, e=35; Br^- ion p=35, n=45, e=36

7.29 a. A potassium ion is smaller than a potassium atom.

b. A bromide ion is larger than a bromine atom.

7.31 A Cl^- ion is larger than a K^+ ion. They are isoelectronic with 18 electrons.

7.33 The Cl^- ion is smaller than S^{2-} ion because it has a greater number of protons.

7.35 Ionization energy - often meaning the first ionization energy- is the amount of energy required to remove an electron from a neutral gaseous atom in its ground state. Yes.

7.37 The first ionization energy increases for elements within each period of elements; the higher the atomic number the higher the ionization energy. Elements in period 3 have lower ionization energies than corresponding elements in period 2.

7.39 Ionization energy within a group decreases as atomic number increases.

7.41 a. Mg has a smaller first ionization energy than S and a greater tendency to form a positive ion. The nuclear charge is greater for a S atom than for a Mg atom.

b. Li has a smaller first ionization energy than F and a greater tendency to form a positive ion. The nuclear charge is greater for a F atom than for a Li atom.

c. Ba has a smaller first ionization energy than Mg and a greater tendency to form a positive ion. The radii of a Ba atom is larger than that of a Mg atom.

7.43 The noble gases have the highest first ionization energies.

7.45 For the alkali metals, melting points decrease as atomic number increases; for the halogens, the trend is the opposite. See Table 7.2.

7.47 a. W b. W c. Cr

7.49 With few exceptions, there is an increase in density for elements in a group (the alkali metals, the halogens, etc.) as atomic number increases.

7.51 a. Al—closer to the center of the period—has a higher density than Mg.

b. Au—closer to the center of the period—has a higher density than Pb.

c. Pt—in the same group but with a higher atomic number—has a higher density than Ni.

7.53 Because hydrogen has one valence electron, it can be placed in Group IA. Because hydrogen, like the halogens, needs one more electron to fill an energy level, it is sometimes placed in Group VIIA. Because properties of hydrogen are not like those of the halogens or of the alkali metals, it is sometimes shown set apart by itself at the top of a periodic table.

7.55 Calcium is the fifth most abundant element in the earth's crust. Calcium carbonate is the chemical present in chalk, limestone, marble, and calcite.

7.57 Aluminum is extracted from bauxite. The extraction process is expensive because the ore undergoes several chemical processes plus an expensive electrolysis process.

7.59 One form of phosphorus is a purplish red noncrystalline form and another is P_4 which is a yellowish white crystalline substance with a waxy appearance.

7.61 C_{60} - buckyballs.

7.63 Oxygen gas, O_2, and ozone, O_3, are both allotropes of oxygen.

7.65 Tin

7.67 Halogen comes from the Greek words meaning "salt former". The halogens react with metals to form salts or ionic compounds.

7.69 When iodine sublimes, it goes directly from solid to the gaseous state without passing through the liquid state.

7.71 Helium was discovered when the astronomer Janssen was using a spectroscope to study an eclipse of the sun. He found a new line in the sun's spectrum.

7.73 Iron is coated with a layer of zinc in the galvanizing process. This retards the rusting of iron.

7.75 The transuranium elements are synthetic elements that follow uranium (atomic number 92) in the periodic table. All transuranium elements are radioactive.

7.77 The P^{3-} ion is the largest of the five ions and decrease in size to the Ca^{2+} ion. They are all isoelectronic with 18 electrons so the element with the greatest number of protons will be the smallest.

7.79 Ionization energies increase as the atomic number increases within a period. The drop in the IE between Be and B is due to the filling of the *s* sublevel and the drop between IE of N and O is due the completion of the half filled *p* sublevel.

7.81 An oxide ion, O^{2-}, is larger than an oxygen atom.

7.83 The second ionization energy of magnesium is higher than the first. It takes more energy to remove each subsequent electron.

7.85 a. Cs has a lower ionization energy than Cl.

b. Zn has a lower ionization energy than Br.

c. Sn has a lower ionization energy than C.

7.87 a. Cs has the greater tendency to form a positive ion. This is because the outermost electron of Cl, the more electronegative element, is held more tightly.

b. Zn has the greater tendency to form a positive ion. This is because the outermost electron of Br, the more electronegative element, is held more tightly.

c. Sn has the greater tendency to form a positive ion. This is because the outermost electron of Sn is farther from the nucleus and thus is more easily removed.

7.89 a. Au - in the same group but with a higher atomic number—has a greater density than Cu.

b. Fe - closer to the center of the period—has a greater density than K.

7.91 Silicon is in quartz, as silicon dioxide, and in carborundum, as silicon carbide. Uses include certain abrasive compounds and computer chips.

7.93 If discovered, element 115 would fall below bismuth in the periodic table. The element would be expected to be a metal. The periodic table allows us to make such predictions because elements in the same group can be expected to have similar chemical properties.

7.95 The calcium ion is smaller. The atom loses 2 e^- to form an ion that is smaller than it's neutral atom.

7.97 Both are rather recently discovered allotropes of carbon.

CHAPTER 8
Chemical Bonds

8.1 Lewis electron-dot symbols: •Mg• → Mg^{2+} + $2e^-$

This is an oxidation reaction. Electrons are lost during oxidation.

8.3 Lewis electron-dot symbols: •S̤̈• + $2e^-$ → :S̤̈:$^{2-}$

This is a reduction reaction. Electrons are gained during reduction.

8.5 •Mg• + •S̤̈• → Mg^{2+} + :S̤̈:$^{2-}$

8.7 An electrolyte is a substance that dissolves in water to give a solution that conducts an electric current. Yes, common table salt, NaCl, is an electrolyte. When the salt dissolves, the dissociated sodium ions and chloride ions in solution allow the solution to conduct electricity.

8.9 An ionic bond is formed when electrons are transferred from one atom to another to form negative and positive ions. A covalent bond is formed when electrons are shared between two atoms.

8.11 Sodium chloride has ionic bonds rather than the covalent bonds present in molecular substances. As a compound, NaCl exists as a lattice of alternating ions, rather than discrete molecules. Only covalently bonded substances exist as molecules.

8.13 In a coordinate covalent bond, one atom donates both of the electrons involved in the shared pair of electrons.

8.15 a. Cl_2 has nonpolar covalent bonds. Both atoms are identical nonmetals.

b. KCl has ionic bonds. Potassium is a metal; chlorine is a nonmetal.

c. The entire CO_2 molecule is linear and nonpolar (Section 8.7). Each C=O bond in CO_2 is polar covalent because the bonds are between different nonmetals.

d. HBr has a polar covalent bond. The electron density is not balanced, so the bond is polar.

e. The entire molecule is nonsymmetrical and polar (Section 8.10). Each N–H bond is polar covalent because the bonds are between different nonmetals.

8.17 a. Br_2 has a nonpolar covalent bond. Both atoms are identical nonmetals.

b. KI has ionic bonds. Potassium is a metal and iodine is a nonmetal.

c. HI has a polar covalent bond. The bonds are between different nonmetals.

d. SO_2 has polar covalent bonds. The bonds are between different nonmetals.

e. PH_3 has nonpolar covalent bonds because phosphorus and hydrogen have the same value of electronegativity. However, the molecule is polar like NH_3, due to its trigional pyramidal shape.

8.19 a. oxidation; loss of e^- b. reduction; gain of e^-

8.21 The molecule is symmetrical with carbon at the center and bromine atoms spaced equidistantly around the central carbon atom.

8.23 $\overset{\delta^+}{H} - \overset{\delta^-}{Br}$

8.25 See Figure 8.6.

a. The difference between Br and Br = 0, so the bond is nonpolar covalent

b. The difference between C and Br = 1.5, so the bond is polar covalent.

c. The difference between Na and Br = 1.9, so the bond is ionic.

8.27 Only $CaCl_2$, compound "c", is an electrolyte; a solution of the compound has ions.

8.29 Nonpolar covalent. (It was a nonconductor and is immiscible with water which is polar, so it must be nonpolar covalent.)

8.31 Metallic. (It was a conductor of electricity as a solid.)

8.33 Battery acid contains ions. It is an electrolyte.

8.35 polar covalent

8.37 a. :C̤̈l· b. :C̤̈l–C̤̈l: c. :C̤̈l:⁻

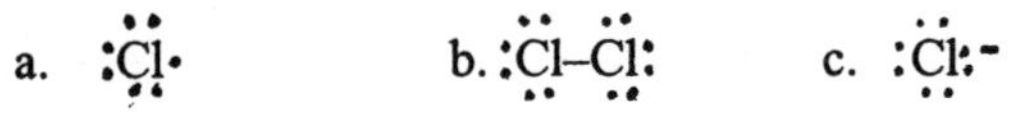

8.39 a

```
   H
   |
H –N:
   |
   H
```

b.

```
     H    +
     |
H – N – H
     |
     H
```

c.

```
      :C̈l:
       |
H – C – C̤̈l:
       |
      :C̤l:
```

d.

```
      H
      |
H – C – Ö – H
      |
      H
```

8.41 a.

```
          :O:
          ||
H – Ö – S – Ö – H
```

b.

```
        :O:    2-
        ||
:Ö –  S – Ö:
```

8.43 :Ö - S̈ = Ö: ↔ :Ö = S̈ - Ö:

8.45 A methane molecule has a tetrahedral shape with 109.5 ° bond angles.

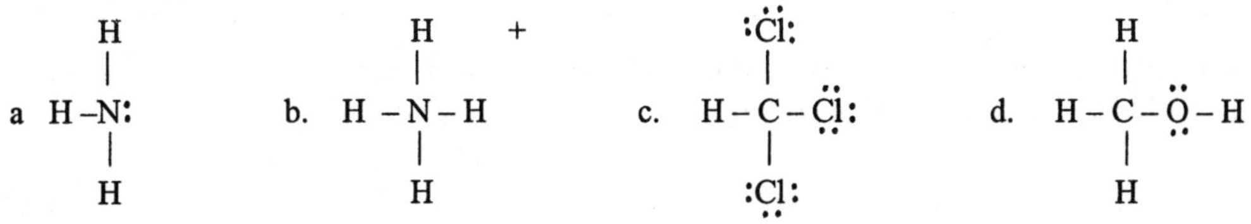

8.47 a.

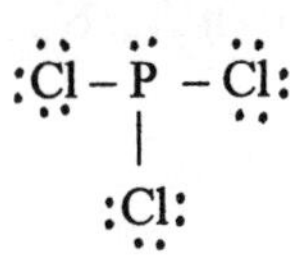

```
:C̤̈l – P̈ – C̤̈l:
       |
     :C̤l:
```

pyramidal shape

b.

```
:C̤̈l – B – C̤̈l:
        |
      :C̤l:
```

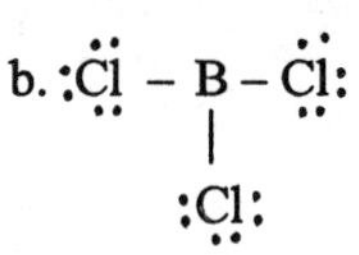

trigonal planar shape

c.

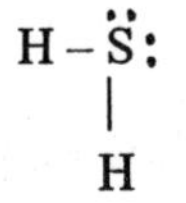

```
H – S̈:
     |
     H
```

bent shape

d. 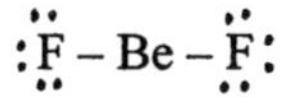 :F̤̈ – Be – F̤̈: linear shape

8.49 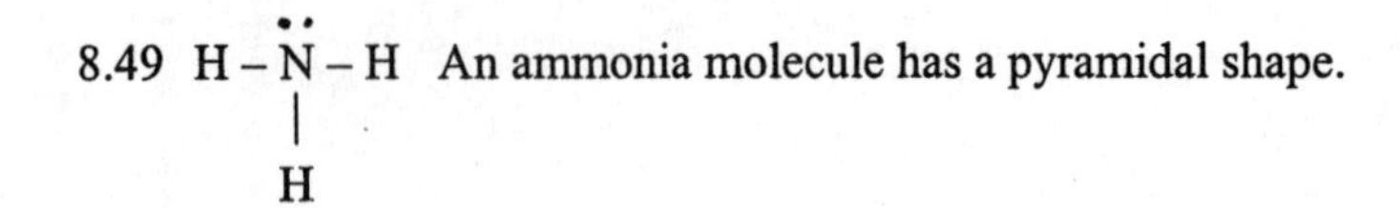
H – N – H An ammonia molecule has a pyramidal shape.

8.51

```
      H  H                          H       H
      |  |   ..                     |  ..   |
    H-C--C-- O -H       and      H- C - O - C -H
      |  |   ..                     |  ..   |
      H  H                          H       H
```

8.53 NI_3 has a pyramidal shape; it is a polar molecule, similar to ammonia, as shown by its Lewis structure.

```
 ..  ..  ..
:I - N - I:
 ..  |   ..
    :I:
     ..
```

8.55 BCl_3 is trigonal planar and CCl_4 is tetrahedral. Both are nonpolar.

8.57 The electrons shared between hydrogen and oxygen atoms within water molecules are all covalent bonds. Hydrogen bonding occurs between water molecules; it is due to the attraction of hydrogen atoms in one molecule for the nonbonding electron pair held by oxygen in another water molecule.

8.59 Hydrogen bonding in water is essential to the weathering of rocks. These attractions cause water to expand as it freezes in the crevices of rocks and other solid materials. When the freezing water expands, it causes these solid materials to crack and break up.

8.61 NF_3 is a pyramidal, polar molecule and BF_3 is a trigonal planar nonpolar molecule. The Lewis structures are

```
 ..  ..  ..            ..      ..
:F - N - F:   and    :F - B - F: .
 ..  |   ..            ..  |   ..
    :F:                   :F:
     ..                    ..
```

Because BF_3 has only three electron pairs, and no nonbonding pairs it has the trigonal planar shape. The extra nonbonding electron pair on N gives NF_3 the polar pyramidal shape.

8.63

```
..   .   ..           ..   .   ..
O  = N - O:    ↔     :O  - N = O
..       ..           ..       ..
```

NO_2 has a bent shape due to the nonbonding electron on N.

8.65 a. An atom of X with 7 valence electrons belongs to Group VIIA. An atom Y with 6 valence electrons belongs to group VIA. An atom of Z with 5 valence electrons belongs to Group VA.

b.

```
 ..           ..           ..
:X - H       :Y - H    H - Z - H
 ..           |            |
              H            H
```

c.

```
       ..               ..
Na+   :X:- ;   2 Na+   :Y: 2-
       ..               ..
```

8.67 Ammonia gas NH_3, is quite polar and dissolves in water, which is also polar, but nitrogen gas, N_2, is nonpolar and is nearly insoluble in water.

8.69

```
          ..                            ..
         :O:                           :O:
          |                             |
    ..         ..                 ..         ..
a. H-O  - S -  O-H        b.  H - O - S -  O - H
    ..    |    ..                 ..  ..   ..
         :O:
          ..
```

8.71
```
  H   H
  |   |
H-N - N-H
  ..  ..
```

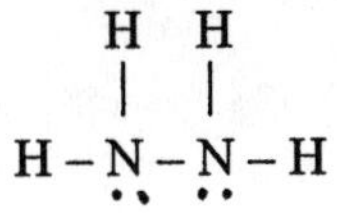

8.73 Ammonia gas molecules are polar and readily dissolve in water, which is also polar, but nonpolar CO_2 molecules are less soluble in water so a carbonated beverage quickly loses its CO_2 gas and the beverage tastes "flat."

8.75 a. no   b. no   c. yes (See Section 8.14.)

8.77 a. no, it is nonpolar.   b. yes, it contains ions.

c. yes, ions dissolve in water.   d. yes, because of hydrogen bonding.

8.79 a. O_2 is nonpolar covalent.   b. K_2O is ionic.   c. CO is polar covalent.

8.81 a. I_2 is nonpolar covalent.   b. HI is polar covalent.   c. HI(aq) is ionic.

8.83 a. 2+   b. 3-   c. 1+   d. 2-

8.85 a. Ar   b. Ar   c. Kr   d. Kr

8.87 a. reduction; electrons are gained.   b. oxidation; electrons are lost.

8.89 nonpolar covalent

8.91 ionic

8.93 a. $\overset{\delta^+}{P} - \overset{\delta^-}{Br}$ b. $\overset{\delta^+}{C} - \overset{\delta^-}{Br}$

8.95 a. $\overset{\delta^+}{H} - \overset{\delta^-}{O}$ b. $\overset{\delta^+}{H} - \overset{\delta^-}{Cl}$

8.97 Both molecules are linear and nonpolar.

8.99 CCl_4 is tetrahedral and nonpolar and will not be soluble in H_2O which is bent and polar.

CHAPTER 9
Chemical Quantities

9.1 The formula weight is an appropriate name for the sum of the atomic masses for any compound. When the compound is molecular—not ionic—the sum can also be called the molecular weight.

9.3 a. 3 Ca = 120.3
2 P = 62.0
8 O = 128.0
F.W. = 310.3

b. 2 C = 24.0
6 H = 6.0
1 O = 16.0
F.W. = 46.0

c. 2 H = 2.0
1 S = 32.1
4 O = 64.0
F.W. = 98.1

d. 1 Mg = 24.3
2 N = 28.0
6 O = 96.0
F.W. = 148.3

9.5 a. 1 mol C = 12.0 g and 2 mol oxygen = 2 x 16.0g = 32.0 g, so 1 mol CO_2 = 44.0 g

b. 1 carbon and 2 oxygen atoms = 3 atoms/molecule

c. $$\frac{3 \text{ atoms}}{1 \text{ molecule}} \times \frac{6.02 \times 10^{23} \text{ molecules}}{1 \text{ mol } CO_2} = \frac{1.81 \times 10^{24} \text{ atoms}}{\text{mol } CO_2}$$

d. $$2.50 \text{ mol } CO_2 \times \frac{44.0 \text{ g } CO_2}{1 \text{ mol } CO_2} = 110. \text{ g } CO_2$$

e. $$70.0 \text{ g } CO_2 \times \frac{1 \text{ mol } CO_2}{44.0 \text{ g } CO_2} = 1.59 \text{ mol } CO_2$$

9.7 a. 6 mol C = 72.0 g, 12 mol H = 12.0 g, and 6 mol O = 96.0 g, so 1 mol $C_6H_{12}O_6$ = 180.0 g

b. 6 C, 12 H, and 6 O atoms = 24 atoms/molecule

c. $$\frac{24 \text{ atoms}}{\text{molecule}} \times \frac{6.02 \times 10^{23} \text{ molecules}}{\text{mol glucose}} = \frac{1.44 \times 10^{24} \text{ atoms}}{\text{mol glucose}}$$

d. $$0.125 \text{ mol glucose} \times \frac{180.0 \text{ g glucose}}{\text{mol glucose}} = 22.5 \text{ g glucose}$$

e. $$50.0 \text{ g glucose} \times \frac{1 \text{ mol glucose}}{180.0 \text{ g glucose}} = 0.278 \text{ mol glucose}$$

9.9 a. 1 Ca^{2+} ions/formula unit of $Ca(OH)_2$

b. 2 OH^- ions/formula unit of $Ca(OH)_2$

c. $$2.50 \text{ mol } Ca(OH)_2 \times \frac{1 \text{ mol } Ca^{2+}}{\text{mol } Ca(OH)_2} = 2.50 \text{ mol } Ca^{2+}$$

d. $$2.50 \text{ mol } Ca(OH)_2 \times \frac{2 \text{ mol } OH^-}{\text{mol } Ca(OH)_2} = 5.00 \text{ mol } OH^-$$

9.11 a. 1 Al^{3+} ions/formula unit of $Al(OH)_3$

b. 3 OH^- ions/formula unit of $Al(OH)_3$

c. $0.222 \text{ mol } Al(OH)_3 \times \frac{1 \text{ mol } Al^{3+}}{\text{mol } Al(OH)_3} = 0.222 \text{ mol } Al^{3+}$

d. $0.222 \text{ mol } Al(OH)_3 \times \frac{3 \text{ mol } OH^-}{\text{mol } Al(OH)_3} = 0.666 \text{ mol } OH^-$

9.13 $1000 \text{ g } (NH_4)_3PO_4 \times \frac{1 \text{ mol } (NH_4)_3PO_4}{149 \text{ g } (NH_4)_3PO_4} \times \frac{3 \text{ mol N}}{1 \text{ mol } (NH_4)_3PO_4} \times \frac{14.0 \text{ g N}}{1 \text{ mol N}} = 282 \text{ g N}$

9.15 $10.0 \text{ g NaCl} \times \frac{1 \text{ mol NaCl}}{58.5 \text{ g NaCl}} \times \frac{1 \text{ mol Na}}{1 \text{ mol NaCl}} \times \frac{23.0 \text{ g } Na^+}{1 \text{ mol Na}} = 3.93 \text{ g } Na^+$

9.17 $10.0 \text{ kg } Zn_2SiO_4 \times \frac{130.8 \text{ g Zn}}{223 \text{ g } Zn_2SiO_4} = 5.87 \text{ kg Zn}$

(This is an alternate method to the one shown for problems 9.13 - 9.15)

9.19 $14.0 \text{ g N}/17.0 \text{ g } NH_3 \times 100\% = 82.4\% \text{ N}$

$3.00 \text{ g H}/17.0 \text{ g } NH_3 \times 100\% = 17.6\% \text{ H}$

9.21 $28.0 \text{ g N}/80.0 \text{ g } NH_4NO_3 \times 100\% = 35.0\% \text{ N}$

$4.0 \text{ g H}/80.0 \text{ g } NH_4NO_3 \times 100\% = 5.0\% \text{ H}$

$48.0 \text{ g O}/80.0 \text{ g } NH_4NO_3 \times 100\% = 60.0\% \text{ O}$

9.23 NH_3, with 82.4% N, has the greatest percentage of N (for compounds given).

9.25 a. $55.8 \text{ g Fe}/151.9 \text{ g } FeSO_4 \times 100\% = 36.7\% \text{ Fe}$

b. $0.367 \times 500. \text{ mg} = 184 \text{ mg Fe}$

9.27 a. $10.0 \text{ g Fe} \times \frac{1 \text{ mol Fe}}{55.8 \text{ g Fe}} = 0.179 \text{ mol Fe}$

b. $10.0 \text{ g } Fe_2O_3 \times \frac{1 \text{ mol } Fe_2O_3}{160. \text{ g } Fe_2O_3} = 0.0625 \text{ mol } Fe_2O_3$

c. $92.0 \text{ g } C_2H_6O \times \frac{1 \text{ mol } C_2H_6O}{46.0 \text{ g } C_2H_6O} = 2.00 \text{ mol } C_2H_6O$

d. $92.0 \text{ g Au} \times \frac{1 \text{ mol Au}}{197.0 \text{ g Au}} = 0.467 \text{ mol Au}$

9.29 a. $0.800 \text{ mol Fe} \times \frac{55.8 \text{ g Fe}}{1 \text{ mol Fe}} = 44.6 \text{ g Fe}$

b. $0.800 \text{ mol } Fe_2O_3 \times \frac{160. \text{ g } Fe_2O_3}{1 \text{ mol } Fe_2O_3} = 128 \text{ g } Fe_2O_3$

c. $1.50 \text{ mol } C_2H_6O \times \dfrac{46.0 \text{ g } C_2H_6O}{1 \text{ mol } C_2H_6O} = 69.0 \text{ g } C_2H_6O$

d. $1.50 \text{ mol Au} \times \dfrac{197.0 \text{ g Au}}{1 \text{ mol Au}} = 296 \text{ g Au}$

9.31 $1 \text{ atom C} \times \dfrac{1 \text{ mol C}}{6.02 \times 10^{23} \text{ atoms C}} \times \dfrac{12.0 \text{ g C}}{1 \text{ mol C}} = 1.99 \times 10^{-23} \text{ g /Catom}$

9.33 $1 \text{ atom N} \times \dfrac{1 \text{ mol N}}{6.02 \times 10^{23} \text{ atoms N}} \times \dfrac{14.0 \text{ g N}}{1 \text{ mol N}} = 2.32 \times 10^{-23} \text{ g/ N atom}$

9.35 $1 \text{ molecule } CO_2 \times \dfrac{1 \text{ mol } CO_2}{6.02 \times 10^{23} \text{ molecules } CO_2} \times \dfrac{44.0 \text{ g } CO_2}{1 \text{ mol } CO_2} = 7.31 \times 10^{-23} \text{ g } CO_2$

9.37 $1 \text{ molecule } C_9H_8O_4 \times \dfrac{1 \text{ mol } C_9H_8O_4}{6.02 \times 10^{23} \text{ molecules } C_9H_8O_4} \times \dfrac{180. \text{ g } C_9H_8O_4}{1 \text{ mol } C_9H_8O_4} = 2.99 \times 10^{-22} \text{ g } C_9H_8O_4$

9.39 $0.0500 \text{ g } H_2O \times \dfrac{1 \text{ mol } H_2O}{18.0 \text{ g } H_2O} \times \dfrac{6.02 \times 10^{23} \text{ molecules}}{1 \text{ mol } H_2O} = 1.67 \times 10^{21} \text{ molecules } H_2O$

9.41

$1.00 \text{ mg Al(OH)}_3 \times \dfrac{1 \text{ g}}{1000 \text{ mg}} \times \dfrac{1 \text{ mol Al(OH)}_3}{78.0 \text{ g Al(OH)}_3} \times \dfrac{3 \text{ mol OH}^- \text{ ions}}{1 \text{ mol Al(OH)}_3} \times \dfrac{6.02 \times 10^{23} \text{ ions}}{1 \text{ mol OH}^- \text{ ions}} =$

$2.32 \times 10^{19} \text{ OH}^- \text{ ions}$

9.43 $1.50 \text{ g } CaCl_2 \times \dfrac{1 \text{ mol } CaCl_2}{111 \text{ g } CaCl_2} \times \dfrac{2 \text{ mol Cl}^- \text{ions}}{1 \text{ mol } CaCl_2} \times \dfrac{6.02 \times 10^{23} \text{ ions}}{1 \text{ mol Cl}^- \text{ions}} = 1.63 \times 10^{22} \text{ Cl}^- \text{ ions}$

9.45

$1.00 \text{ mg } C_6H_8O_6 \times \dfrac{1 \text{ g}}{1000 \text{ mg}} \times \dfrac{1 \text{ mol } C_6H_8O_6}{176 \text{ g } C_6H_8O_6} \times \dfrac{6.02 \times 10^{23} \text{ molecules}}{1 \text{ mol } C_6H_8O_6} =$

$3.42 \times 10^{18} \text{ } C_6H_8O_6 \text{ molecules}$

9.47 $0.500 \text{ L} \times \dfrac{0.100 \text{ mol } KNO_3}{1 \text{ L}} \times \dfrac{101.1 \text{ g } KNO_3}{1 \text{ mol } KNO_3} = 5.06 \text{ g } KNO_3$

Dissolve the compound in enough water to make 500 mL of solution.

9.49 $0.250\ L \times \frac{0.150\ mol\ C_6H_{12}O_6}{1\ L} \times \frac{180.\ g\ C_6H_{12}O_6}{1\ mol\ C_6H_{12}O_6} = 6.75\ g\ C_6H_{12}O_6$

Dissolve the compound in enough water to make 250 mL of solution.

9.51 $2.00\ g\ C_6H_{12}O_6 \times \frac{1\ mol\ C_6H_{12}O_6}{180.\ g\ C_6H_{12}O_6} \times \frac{1\ L}{0.150\ mol} \times \frac{1000\ mL}{1\ L} = 74.1\ mL$

9.53 $0.500\ g\ C_6H_{12}O_6 \times \frac{1\ mol\ C_6H_{12}O_6}{180.\ g\ C_6H_{12}O_6} \times \frac{1\ L}{0.150\ mol} \times \frac{1000\ mL}{1\ L} = 18.5\ mL$

9.55 $V_1C_1 = V_2C_2$ $\quad V_1 \times 12.0\ M = 2000.\ mL \times 0.100\ M$ $\quad V_1 = \frac{2000.\ mL \times 0.100\ M}{12.0\ M} = 16.7\ mL$

9.57 $V_1C_1 = V_2C_2$ $\quad V_1 \times 5.25\% = 50.0\ L \times 1.00\%$ $\quad V_1 = \frac{50.0\ L \times 1.00\%}{5.25\%} = 9.52\ L$

9.59 a. Half of $C_8H_{18} = C_4H_9$ $\quad$ b. $C_{12}H_{22}O_{11}$

c. Half of $Hg_2Cl_2 = HgCl$ $\quad$ d. $CaCl_2$

9.61 a. $\frac{\text{Actual F.W.}}{\text{Empirical F.W.}} = \frac{84.0}{14.0} = 6$ units $\quad 6 \times CH_2 = C_6H_{12}$

b. $\frac{\text{Actual F.W.}}{\text{Empirical F.W.}} = \frac{60.0}{30.0} = 2$ units $\quad 2 \times CH_2O = C_2H_4O_2$

c. $\frac{\text{Actual F.W.}}{\text{Empirical F.W.}} = \frac{176.0}{88.0} = 2$ units $\quad 2 \times C_3H_4O_3 = C_6H_8O_6$

d. $\frac{\text{Actual F.W.}}{\text{Empirical F.W.}} = \frac{27.7}{13.8} = 2$ units $\quad 2 \times BH_3 = B_2H_6$

9.63

P: $0.0186\ g\ P \times \frac{1\ mol\ P}{31.0\ g\ P} = 0.000600\ mol\ P$ $\quad 0.000600 / 0.000600 = 1.00 \times 2 = 2$

N: $0.0126\ g\ N \times \frac{1\ mol\ N}{14.0\ g} = 0.000900\ mol\ N$ $\quad 0.000900 / 0.000600 = 1.50 \times 2 = 3$

Empirical formula: P_2N_3

9.65

P: $35.6\ g\ P \times \frac{1\ mol\ P}{31.0\ g\ P} = 1.15\ mol\ P$ $\quad 1.15 / 1.15 = 1.00 \times 4 = 4$

N: $64.4\ g\ S \times \frac{1\ mol\ S}{32.1\ g} = 2.01\ mol\ S$ $\quad 2.01 / 1.15 = 1.75 \times 4 = 7$

Empirical formula: P_4S_7

9.67

$$N : 36.84 \text{ g N x } \frac{1 \text{ mol N}}{14.01 \text{ g N}} = 2.631 \text{ mol N} \qquad 2.631/2.631 = 1.000 \text{ x } 2 = 2$$

$$O : 63.16 \text{ g O x } \frac{1 \text{ mol O}}{16.00 \text{ g O}} = 3.944 \text{ mol O} \qquad 3.944/2.631 = 1.499 \text{ x } 2 = 3$$

Empirical formula : N_2O_3

9.69

$$C\text{: } 67.89 \text{ g C x } \frac{1 \text{ mol C}}{12.0 \text{ g C}} = 5.658 \text{ mol C} \qquad 5.657 / 1.886 = 3$$

$$H\text{: } 5.71 \text{ g H x } \frac{1 \text{ mol H}}{1.01 \text{ g}} = 5.65 \text{ mol H} \qquad 5.65 / 1.886 = 3$$

$$N\text{: } 26.40 \text{ g N x } \frac{1 \text{ mol N}}{14.00 \text{ g N}} = 1.886 \text{ mol N} \qquad 1.886 / 1.886 = 1$$

Empirical formula: C_3H_3N

9.71 a.

$$H\text{: } 12.5 \text{ g H x } \frac{1 \text{ mol H}}{1.01 \text{ g}} = 12.4 \text{ mol H} \qquad 12.4 / 6.25 = 2$$

$$N\text{: } 87.5 \text{ g N x } \frac{1 \text{ mol N}}{14.0 \text{ g N}} = 6.25 \text{ mol N} \qquad 6.25 / 6.25 = 1$$

Empirical formula: NH_2

b. $\frac{\text{Actual F.W.}}{\text{Empirical F.W.}} = \frac{32.0}{16.0} = 2 \text{ units} \qquad 2 \text{ x } NH_2 = N_2H_4$

9.73 a.

$$C\text{: } 24.49 \text{ g C x } \frac{1 \text{ mol C}}{12.0 \text{ g C}} = 2.04 \text{ mol C} \qquad 2.04 / 2.04 = 1$$

$$H\text{: } 4.08 \text{ g H x } \frac{1 \text{ mol H}}{1.01 \text{ g}} = 4.04 \text{ mol H} \qquad 4.04 / 2.04 = 2$$

$$Cl\text{: } 72.43 \text{ g Cl x } \frac{1 \text{ mol Cl}}{35.5 \text{ g Cl}} = 2.04 \text{ mol Cl} \qquad 2.04 / 2.04 = 1$$

Empirical formula: CH_2Cl

b. $\frac{\text{Actual F.W.}}{\text{Empirical F.W.}} = \frac{98.0}{49.5} = 2 \text{ units} \qquad 2 \text{ x } CH_2Cl = C_2H_4Cl_2$

9.75

$$600. \text{ mg } Li_2CO_3 \text{ x } \frac{1 \text{ g}}{1000 \text{ mg}} \text{ x } \frac{1 \text{ mol } Li_2CO_3}{73.8 \text{ g } Li_2CO_3} \text{ x } \frac{2 \text{ mol } Li^+ \text{ ions}}{1 \text{ mol } Li_2CO_3} \text{ x } \frac{6.02 \text{ x } 10^{23} \text{ } Li^+ \text{ ions}}{1 \text{ mol } Li^+ \text{ ions}} =$$

$9.79 \text{ x } 10^{21} \text{ } Li^+ \text{ ions}$

9.77

$$0.0120\,L \times 0.100\,M = 2.00\,L \times C_2 \quad C_2 = 0.000600M \quad 0.000100\,L \times \frac{0.000600\text{ mol }CaCl_2}{1\,L} \times$$

$$\frac{2\text{ mol }Cl^-\text{ ions}}{1\text{ mol }CaCl_2} \times \frac{6.02 \times 10^{23}\ Cl^-\text{ ions}}{1\text{ mol }Cl^-\text{ ions}} = 7.22 \times 10^{16}\ Cl^-\text{ ions}$$

9.79 1 mol H_2O = 2 mol H + 1 mol O = 2.0 g + 16.0 g = 18.0 g/mol H_2O

$$1\text{ molecule }H_2O \times \frac{1\text{ mol }H_2O}{6.02 \times 10^{23}\text{ molecules }H_2O} \times \frac{18.0\text{ g }H_2O}{1\text{ mol }H_2O} = 2.99 \times 10^{-23}\text{ g }H_2O$$

9.81 1 mol $(NH_4)_3PO_4$ = 3 mol N = 42.0 g N; 12 mol H = 12.0 g H; 1 mol P = 31.0 g P; 4 mol O = 64.0 g O;
so 1 mol of $(NH_4)_3PO_4$ = 149 g

$$0.100\text{ g }(NH_4)_3PO_4 \times \frac{1\text{ mol }(NH_4)_3PO_4}{149\text{ g }(NH_4)_3PO_4} \times \frac{3\text{ mol }NH_4^+\text{ ions}}{1\text{ mol }(NH_4)_3PO_4} \times \frac{6.02 \times 10^{23}\text{ ions}}{1\text{ mol }NH_4^+\text{ ions}} =$$

$1.21 \times 10^{21}\ NH_4^+$ ions

9.83 $V_1C_1 = V_2C_2$ $\quad V_1 \times 10.0\% = 250\text{ mL} \times 2.0\%$ $\quad V_1 = \frac{250.\text{ mL} \times 2.0\%}{10.0\%} = 50.\text{ mL}$

9.85 $1.00\,L \times \frac{0.200\text{ mol }CuSO_4 \bullet 5H_2O}{1\,L} \times \frac{250.\text{ g }CuSO_4 \bullet 5H_2O}{1\text{ mol }CuSO_4 \bullet 5H_2O} = 50.0\text{ g }CuSO_4 \bullet 5H_2O$

9.87 $6.00 \times 10^{-3}\text{ mol }Cu^{2+} \times \frac{1\text{ mol }CuSO_4}{1\text{ mol }Cu^{2+}\text{ ions}} \times \frac{1\text{ L solution}}{0.200\text{ mol }CuSO_4} \times \frac{1000\text{ mL}}{1\,L} = 30.0\text{ mL}$

9.89 $1.00 \times 10^{-3}\,L \times \frac{0.100\text{ mol }CaCl_2}{1\,L} \times \frac{2\text{ mol }Cl^-\text{ ions}}{1\text{ mol }CaCl_2} \times \frac{6.02 \times 10^{23}\ Cl^-\text{ ions}}{1\text{ mol }Cl^-\text{ ions}} = 1.20 \times 10^{20}\ Cl^-\text{ ions}$

9.91 $50.0\text{ g glycine} \times \frac{1\text{ mol glycine}}{75.0\text{ g glycine}} \times \frac{1\text{ mol N}}{1\text{ mol glycine}} \times \frac{14.0\text{ g N}}{1\text{ mol N}} = 9.33\text{ g N}$

9.93 $30.0\text{ g sucrose} \times \frac{1\text{ mol sucrose}}{342\text{ g sucrose}} \times \frac{6.02 \times 10^{23}\text{ molecules}}{1\text{ mol sucrose}} = 5.28 \times 10^{22}\text{ molecules sucrose}$

9.95 $15.0\text{ mL} \times \frac{1\,L}{1000\text{ mL}} \times \frac{0.200\text{ mol }NaHCO_3}{1\,L} \times \frac{84.0\text{ g }NaHCO_3}{1\text{ mol }NaHCO_3} = 0.252\text{ g }NaHCO_3$

9.97 $V_1 \times 6.0\text{ M} = 500\text{ mL} \times 0.100\text{ M}$ $\quad V_1 = \frac{500.\text{ mL} \times 0.100\text{ M}}{6.00\text{ M}} = 8.33\text{ mL}$

9.99 $12.0 \text{ mL} \times \frac{1 \text{ L}}{1000 \text{ mL}} \times \frac{0.100 \text{ mol NaOH}}{1 \text{ L}} \times \frac{40.0 \text{ g NaOH}}{1 \text{ mol NaOH}} \times \frac{1000 \text{ mg}}{1 \text{ g}} = 48.0 \text{ mg NaOH}$

9.101

C: $82.62 \text{ g C} \times \frac{1 \text{ mol C}}{12.0 \text{ g C}} = 6.885 \text{ mol C}$ $6.885 / 6.885 = 1 \times 2 = 2$

H: $17.38 \text{ g H} \times \frac{1 \text{ mol H}}{1.01 \text{ g}} = 17.21 \text{ mol H}$ $17.21 / 6.885 = 2.5 \times 2 = 5$

Empirical formula: C_2H_5

$\frac{\text{Actual F.W.}}{\text{Empirical F.W.}} = \frac{58.1}{29.0} = 2 \text{ units}$ $2 \times C_2H_5 = C_4H_{10}$

9.103

Na : $17.56 \text{ g Na} \times \frac{1 \text{ mol Na}}{22.99 \text{ g Na}} = 0.7638 \text{ mol Na}$ $0.7638/0.7632 = 1 \times 2 = 2$

Cr : $39.69 \text{ g Cr} \times \frac{1 \text{ mol Cr}}{52.00 \text{ g Cr}} = 0.7632 \text{ mol Cr}$ $0.7632/0.7632 = 1 \times 2 = 2$

O : $42.75 \text{ g O} \times \frac{1 \text{ mol O}}{16.00 \text{ g O}} = 2.672 \text{ mol O}$ $2.672/0.7632 = 3.501 \times 2 = 7$

Empirical formula : $Na_2Cr_2O_7$

9.105

N : $55.26 \text{ mg N} \times \frac{1 \text{ mmol N}}{14.01 \text{ mg N}} = 3.944 \text{ mmol N}$ $3.944/3.944 = 1 \times 2 = 2$

O : $94.74 \text{ mg O} \times \frac{1 \text{ mmol O}}{16.00 \text{ mg O}} = 5.921 \text{ mmol O}$ $5.921/3.944 = 1.501 \times 2 = 3$

Empirical formula : N_2O_3

9.107

N : $25.93 \text{ g N} \times \frac{1 \text{ mol N}}{14.01 \text{ g N}} = 1.851 \text{ mol N}$ $1.851/1.851 = 1 \times 2 = 2$

O : $74.07 \text{ g O} \times \frac{1 \text{ mol O}}{16.00 \text{ g O}} = 4.626 \text{ mol O}$ $4.626/1.851 = 2.501 \times 2 = 5$

Empirical formula : N_2O_5 $(2 \times 14.01) + (5 \times 16.00) = 108$

Molecular Formula : N_2O_5

9.109 $650. \text{ mg CaCO}_3 \times \frac{40.1 \text{ mg Ca}^{2+}}{100.1 \text{ mg CaCO}_3} = 260. \text{ mg Ca}^{2+}$

CHAPTER 10
Chemical Reactions

10.1 In a balanced chemical equation the total masses of reactants and products are equal. This is in agreement with the law of conservation of mass.

10.3 Atoms and mass. Molecules and moles do not need to balance since they contain varying number of atoms; if atoms are balanced then mass will always be balanced.

10.5 a. $4\,Cr + 3\,O_2 \rightarrow 2Cr_2O_3$

b. $2\,SO_2 + O_2 \rightarrow 2\,SO_3$

c. $2\,PbO_2 \rightarrow 2\,PbO + O_2(g)$

d. $2\,NaOH + CO_2 \rightarrow Na_2CO_3 + H_2O$

e. $Al_2(SO_4)_3 + 6\,NaOH \rightarrow 2\,Al(OH)_3 + 3\,Na_2SO_4$

10.7 Choices a, b, and c are true. Choices d and e are false.

10.9 a. $Mg + 2\,H_2O \rightarrow Mg(OH)_2 + H_2(g)$

b. $2\,NaHCO_3 + H_3PO_4 \rightarrow Na_2HPO_4 + 2\,H_2O + 2\,CO_2$

c. $2\,Al + 3\,H_2SO_4(aq) \rightarrow Al_2(SO_4)_3 + 3\,H_2(g)$

d. $C_3H_8 + 5\,O_2(g) \rightarrow 3\,CO_2 + 4\,H_2O$

e. $2\,CH_3OH + 3\,O_2(g) \rightarrow 2\,CO_2 + 4\,H_2O$

10.11 a. $CaCO_3 + 2\,HCl \rightarrow CaCl_2 + H_2O + CO_2$

b. $PCl_5 + 4\,H_2O \rightarrow H_3PO_4 + 5\,HCl$

c. $2\,KClO_3 \rightarrow 2\,KCl + 3\,O_2(g)$

d. $3\,Ba(OH)_2 + 2\,H_3PO_4 \rightarrow Ba_3(PO_4)_2 + 6\,H_2O$

e. $2\,C_2H_6(g) + 7\,O_2(g) \rightarrow 4\,CO_2(g) + 6\,H_2O(g)$

10.13 a. decomposition b. single replacement c. combustion d. double replacement
e. decomposition

10.15 combination

10.17 $2\,NaBr(aq) + Cl_2(g) \rightarrow 2\,NaCl(aq) + Br_2(l)$ single replacement

10.19 $2\,C_4H_{10} + 13\,O_2(g) \rightarrow 8\,CO_2 + 10\,H_2O$

10.21 $C_3H_6O + 4\,O_2(g) \rightarrow 3\,CO_2 + 3\,H_2O$

10.23 $C_3H_8 + 5\,O_2(g) \rightarrow 3\,CO_2 + 4\,H_2O$

10.25 a. $2\,CO(g) + O_2(g) \rightarrow 2\,CO_2$

b. $8\,Zn + S_8 \rightarrow 8\,ZnS$

c. $N_2(g) + 3\,H_2(g) \rightarrow 2\,NH_3(g)$

d. $2\,NO(g) + O_2(g) \rightarrow 2\,NO_2(g)$

e. $2\,Fe + 3\,Br_2 \rightarrow 2\,FeBr_3$

10.27 a. $2\,Al_2O_3 \rightarrow 4\,Al + 3\,O_2(g)$

b. $2\,PbO_2 \rightarrow 2\,PbO + O_2(g)$

c. $2\,NaClO_3 \rightarrow 2\,NaCl + 3\,O_2(g)$

d. $2\,KNO_3 \rightarrow 2\,KNO_2 + O_2(g)$

e. $2\,H_2O_2 \rightarrow 2\,H_2O + O_2(g)$

10.29 a. $4\,Li + O_2 \rightarrow 2\,Li_2O$

b. $2\,Al + 3\,I_2 \rightarrow 2\,AlI_3$

c. $2\,Fe + 3\,Br_2 \rightarrow 2\,FeBr_3$

d. $4\,Al + 3\,O_2 \rightarrow 2\,Al_2O_3$

10.31 $4\,Cu + O_2 \rightarrow 2\,Cu_2O$

10.33 a. $Li_2O + H_2O \rightarrow 2\,LiOH$

b. $2\,Na + 2\,H_2O \rightarrow 2\,NaOH + H_2(g)$

c. $Mg + 2\,H_2O(g) \rightarrow Mg(OH)_2 + H_2(g)$

d. $Ag + H_2O \rightarrow$ no reaction

e. $SrO + H_2O \rightarrow Sr(OH)_2$

10.35 a. $Zn + 2\,HCl\,(aq) \rightarrow ZnCl_2(aq) + H_2(g)$

b. $Cu + HCl(aq) \rightarrow$ no reaction

c. $Mg + Fe(NO_3)_2(aq) \rightarrow Mg(NO_3)_2(aq) + Fe$

d. $3\,AgNO_3(aq) + Al \rightarrow Al(NO_3)_3(aq) + 3\,Ag$

e. $Fe + MgCl_2(aq) \rightarrow$ no reaction

10.37 a. $2\ Al + 3\ H_2SO_4(aq) \rightarrow Al_2(SO_4)_3(aq) + 3\ H_2(g)$

b. $Au + HCl(aq) \rightarrow$ no reaction

c. $Fe + CuSO_4(aq) \rightarrow FeSO_4(aq) + Cu$ The equation is balanced for iron(II) sulfate, but iron(III) sulfate could also be produced.

d. Acetic acid (present in apples) does not react with copper.
$CH_3COOH + Cu \rightarrow$ no reaction

e. Yes. Lead (II) ions react with aluminum metal.
$3\ Pb(NO_3)_2(aq) + 2\ Al \rightarrow 2\ Al(NO_3)_3(aq) + 3\ Pb$

10.39 a. $S_8 + 8\ O_2(g) \rightarrow 8\ SO_2(g)$

b. $SO_2 + H_2O \rightarrow H_2SO_3$

c. $N_2O_5 + H_2O \rightarrow 2\ HNO_3$

d. $2\ KBr + Cl_2 \rightarrow 2\ KCl + Br_2$

e. $KCl + I_2 \rightarrow$ no reaction

10.41 The chlorine would react with HI gas to produce elemental iodine, I_2.

$Cl_2 + 2\ HI \rightarrow 2\ HCl + I_2$

10.43 a. $AgNO_3(aq) + KCl(aq) \rightarrow \underline{AgCl}(s) + KNO_3(aq)$

b. $FeCl_3(aq) + 3\ NaOH(aq) \rightarrow \underline{Fe(OH)_3}(s) + 3\ NaCl(aq)$

c. $Al_2(SO_4)_3(aq) + 3\ Ba(NO_3)_2(aq) \rightarrow 2\ Al(NO_3)_3(aq) + 3\ \underline{BaSO_4}(s)$

d. $Pb(NO_3)_2(aq) + K_2Cr_2O_7(aq) \rightarrow \underline{PbCr_2O_7}(s) + 2\ KNO_3(aq)$

e. $2\ AgNO_3(aq) + K_2CrO_4(aq) \rightarrow \underline{Ag_2CrO_4}(s) + 2\ KNO_3(aq)$

10.45 a. $NH_4Cl(aq) + KNO_3(aq) \rightarrow$ no reaction a precipitate is not formed

b. $CaCO_3(s) + 2\ HCl(aq) \rightarrow CaCl_2(aq) + H_2O + CO_2(g)$

c. $2\ HCl(aq) + Na_2S(aq) \rightarrow 2\ NaCl + H_2S(g)$

d. $BaCl_2(aq) + K_2SO_4(aq) \rightarrow BaSO_4(s) + 2\ KCl(aq)$

10.47 a. $Ag^+(aq) + NO_3^-(aq) + K^+(aq) + Cl^-(aq) \rightarrow AgCl(s) + K^+(aq) + NO_3^-(aq)$

b. $Fe^{3+}(aq) + 3\ Cl^-(aq) + 3\ Na^+(aq) + 3\ OH^-(aq) \rightarrow Fe(OH)_3(s) + 3\ Na^+(aq) + Cl^-(aq)$

c. $2\ Al^{3+}(aq) + 3\ SO_4^{2-}(aq) + 3\ Ba^{2+}(aq) + 6\ NO_3^-(aq) \rightarrow 2\ Al^{3+}(aq) + 6\ NO_3^-(aq) + 3\ BaSO_4(s)$

d. $Pb^{2+}(aq) + 2\ NO_3^-(aq) + 2\ K^+(aq) + Cr_2O_7^{2-}(aq) \rightarrow PbCr_2O_7(s) + 2\ K^+(aq) + 2\ NO_3^-(aq)$

e. $2\ Ag^+(aq) + 2\ NO_3^-(aq) + 2\ K^+(aq) + CrO_4^{2-}(aq) \rightarrow Ag_2CrO_4(s) + 2\ K^+(aq) + 2\ NO_3^-(aq)$

10.49 a. $Ag^+(aq) + Cl^-(aq) \rightarrow AgCl(s)$

b. $Fe^{3+}(aq) + 3\ OH^-(aq) \rightarrow Fe(OH)_3(s)$

c. $Ba^{2+}(aq) + SO_4^{2-}(aq) + \rightarrow BaSO_4(s)$

d. $Pb^{2+}(aq) + Cr_2O_7^{2-}(aq) \rightarrow PbCr_2O_7(s)$

e. $2\ Ag^+(aq) + CrO_4^{2-}(aq) \rightarrow Ag_2CrO_4(s)$

10.51 $2\ AgNO_3(aq) + CaCl_2(aq) \rightarrow 2\ AgCl(s) + Ca(NO_3)_2(aq)$

$2\ Ag^+(aq) + 2\ NO_3^-(aq) + Ca^{2+}(aq) + 2Cl^-(aq) \rightarrow 2\ AgCl(s) + Ca^{2+}(aq) + 2\ NO_3^-(aq)$

$Ag^+(aq) + Cl^-(aq) \rightarrow AgCl(s)$

10.53 A hydrogen ion, H^+, combines with hydroxide ion, OH^-, to produce water.

10.55 a. $H_2SO_4(aq) + 2\ KOH(aq) \rightarrow K_2SO_4(aq) + 2\ H_2O$

b. acid base salt water

c. $H^+(aq) + OH^-(aq) \rightarrow H_2O$

d. K^+, SO_4^{2-}

10.57 The acetic acid present in apples reacted with the iron.

$Fe + 2CH_3COOH(aq) \rightarrow Fe(CH_3COO)_2(aq) + H_2(g)$

10.59 Double replacement reactions.

10.61 $Cu + 2\ H_2SO_4 \rightarrow SO_2 + CuSO_4 + 2\ H_2O$

10.63 $2\ HCl + MgCO_3 \rightarrow MgCl_2 + H_2O + CO_2(g)$

10.65 double replacement

10.67 synthesis

10.69 $Fe_2O_3 + 3\ CO(g) \rightarrow 2\ Fe + 3\ CO_2(g)$

10.71 $2\ HCl + CaCO_3 \rightarrow CaCl_2 + H_2O + CO_2(g)$

10.73 $Ca(OH)_2 + CO_2 \rightarrow CaCO_3 + H_2O$

10.75 $6\ NaN_3 + Fe_2O_3 \rightarrow 3\ Na_2O + 2\ Fe + 9\ N_2\ (g)$

10.77 $2\ Al + 6\ HCl \rightarrow 2\ AlCl_3 + 3\ H_2$

10.79 $2\ Al_2O_3 + 3\ C \rightarrow 3\ CO_2 + 4\ Al$

10.81 a. $3\ H_2SO_4 + 2\ Al(OH)_3 \rightarrow Al_2(SO_4)_3 + 6\ H_2O$

b. $2\ PbO_2 \rightarrow 2\ PbO + O_2$

c. a. double-replacement and neutralization b. decomposition and oxidation-reduction

10.83 a. $3\ Fe(NO_3)_2 + 2\ Al \rightarrow 2\ Al(NO_3)_3 + 3\ Fe$

b. $Al_2(SO_4)_3 + 3\ Pb(NO_3)_2 \rightarrow 2\ Al(NO_3)_3 + 3\ PbSO_4$

c. a. single-replacement and oxidation-reduction b. double-replacement

10.85 a. $2\ Li + 2\ H_2O \rightarrow 2\ LiOH + H_2$

b. $Cl_2 + SnF_2 \rightarrow$ no reaction

c. a. single replacement and oxidation-reduction

10.87 a. $Cu + Cl_2 \rightarrow CuCl_2$ b. chlorine c. copper

10.89 a. $Fe_2O_3 + 3\ CO \rightarrow 2\ Fe + 3\ CO_2$ b. CO c. iron(III)

10.91 a. $Ca(OH)_2 + 2\ HNO_3 \rightarrow Ca(NO_3)_2 + 2\ H_2O$

b. acid: HNO_3; base: $Ca(OH)_2$; salt: $Ca(NO_3)_2$

c. $H^+ + OH^- \rightarrow H_2O$ d. Ca^{2+}, NO_3^-

10.93 $CaCO_3\ (s) + 2\ HCl\ (aq) \rightarrow CaCl_2\ (aq) + CO_2\ (g) + H_2O\ (l)$

10.95 a. Zn d. Mg

10.97 $MgCO_3 \rightarrow MgO + CO_2$

10.99 $2\ NH_4NO_3 \rightarrow 2\ N_2 + O_2 + 4\ H_2O$

CHAPTER 11
Stoichiometry: Calculations Based on Chemical Equations

11.1 a. $\frac{2 \text{ mol } HNO_3}{3 \text{ mol } NO_2}$ b. $\frac{1 \text{ mol NO}}{3 \text{ mol } NO_2}$

c. $\frac{1 \text{ mol } H_2O}{2 \text{ mol } HNO_3}$ d. $\frac{2 \text{ mol } HNO_3}{1 \text{ mol NO}}$

11.3 a. $63.3 \text{ mol } NO_2 \times \frac{2 \text{ mol } HNO_3}{3 \text{ mol } NO_2} = 42.2 \text{ mol } HNO_3$

b. $12.3 \text{ mol } NO_2 \times \frac{1 \text{ mol NO}}{3 \text{ mol } NO_2} = 4.10 \text{ mol NO}$

c. $6.44 \text{ mol } HNO_3 \times \frac{1 \text{ mol } H_2O}{2 \text{ mol } HNO_3} = 3.22 \text{ mol } H_2O$

d. $7.25 \text{ mol NO} \times \frac{2 \text{ mol } HNO_3}{1 \text{ mol NO}} = 14.5 \text{ mol } HNO_3$

11.5 a. $0.684 \text{ mol HCl} \times \frac{1 \text{ mol } Ca(OH)_2}{2 \text{ mol HCl}} \times \frac{74.0 \text{ g } Ca(OH)_2}{1 \text{ mol } Ca(OH)_2} = 25.3 \text{ g } Ca(OH)_2$

b. $0.684 \text{ mol HCl} \times \frac{2 \text{ mol } H_2O}{2 \text{ mol HCl}} \times \frac{18.0 \text{ g } H_2O}{1 \text{ mol } H_2O} = 12.3 \text{ g } H_2O$

c. $0.684 \text{ mol HCl} \times \frac{1 \text{ mol } CaCl_2}{2 \text{ mol HCl}} \times \frac{111.0 \text{ g } CaCl_2}{1 \text{ mol } CaCl_2} = 38.0 \text{ g } CaCl_2$

d. $0.684 \text{ mol HCl} \times \frac{36.5 \text{ g HCl}}{1 \text{ mol HCl}} = 25.0 \text{ g HCl}$

e. There are 50.3 g of reactants and 50.3 g of products for the reaction. Mass is conserved; matter is neither created nor destroyed.

11.7 a. $50.0 \text{ g } C_8H_{18} \times \frac{1 \text{ mol } C_8H_{18}}{114.0 \text{ g } C_8H_{18}} \times \frac{25 \text{ mol } O_2}{2 \text{ mol } C_8H_{18}} = 5.48 \text{ mol } O_2$

b. $50.0 \text{ g } C_8H_{18} \times \frac{1 \text{ mol } C_8H_{18}}{114.0 \text{ g } C_8H_{18}} \times \frac{25 \text{ mol } O_2}{2 \text{ mol } C_8H_{18}} \times \frac{32.0 \text{ g } O_2}{1 \text{ mol } O_2} = 175 \text{ g } O_2$

c. $50.0 \text{ g } C_8H_{18} \times \frac{1 \text{ mol } C_8H_{18}}{114.0 \text{ g } C_8H_{18}} \times \frac{16 \text{ mol } CO_2}{2 \text{ mol } C_8H_{18}} = 3.51 \text{ mol } CO_2$

d. $50.0\ g\ C_8H_{18} \times \frac{1\ mol\ C_8H_{18}}{114.0\ g\ C_8H_{18}} \times \frac{16\ mol\ CO_2}{2\ mol\ C_8H_{18}} \times \frac{44.0\ g\ CO_2}{1\ mol\ CO_2} = 154\ g\ CO_2$

11.9 $2\ NaN_3 \rightarrow 2\ Na + 3\ N_2$

$$450.\ g\ NaN_3 \times \frac{1\ mol\ NaN_3}{65.0\ g\ NaN_3} \times \frac{3\ mol\ N_2}{2\ mol\ NaN_3} \times \frac{28.0\ g\ N_2}{1\ mol\ N_2} = 291\ g\ N_2$$

11.11 $CaCO_3 + 2\ HCl \rightarrow CaCl_2 + CO_2 + H_2O$

$$1.000\ g\ CaCO_3 \times \frac{1\ mol\ CaCO_3}{100.1\ g\ CaCO_3} \times \frac{2\ mol\ HCl}{1\ mol\ CaCO_3} \times \frac{36.46\ g\ HCl}{1\ mol\ HCl} = 0.7285\ g\ HCl$$

11.13 a. $Mg + 2\ HCl\ (aq) \rightarrow MgCl_2 + H_2$

b. $20.0\ g\ Mg \times \frac{1\ mol\ Mg}{24.3\ g\ Mg} \times \frac{1\ mol\ H_2}{1\ mol\ Mg} = 0.823\ mol\ H_2$

c. $20.0\ g\ Mg \times \frac{1\ mol\ Mg}{24.3\ g\ Mg} \times \frac{2\ mol\ HCl}{1\ mol\ Mg} \times \frac{36.5\ g\ HCl}{1\ mol\ HCl} = 60.1\ g\ HCl$

11.15 $20.0\ g\ Mg \times \frac{1\ mol\ Mg}{24.3\ g\ Mg} \times \frac{2\ mol\ HCl}{1\ mol\ Mg} \times \frac{1.00\ L\ acid}{6.00\ mol\ HCl} \times \frac{1000\ mL}{1\ L} = 274\ mL\ acid$

11.17 $Ca(OH)_2 + 2\ HCl\ (aq) \rightarrow CaCl_2 + 2\ H_2O$

$$10.0\ g\ Ca(OH)_2 \times \frac{1\ mol\ Ca(OH)_2}{74.1\ g\ Ca(OH)_2} \times \frac{2\ mol\ HCl}{1\ mol\ Ca(OH)_2} \times \frac{1.00\ L\ acid}{3.00\ mol\ HCl} \times \frac{1000\ mL}{1.00\ L} = 90.0\ mL\ acid$$

11.19 $CaCO_3 + 2\ HCl \rightarrow CaCl_2 + CO_2 + H_2O$

$$0.500\ g\ CaCO_3 \times \frac{1\ mol\ CaCO_3}{100.1\ g\ CaCO_3} \times \frac{2\ mol\ HCl}{1\ mol\ CaCO_3} \times \frac{1.00\ L\ acid}{0.100\ mol\ HCl} \times \frac{1000\ mL}{1\ L} = 100.\ mL\ acid$$

11.21 a. $4.00\ g\ Al \times \frac{1\ mol\ Al}{27.0\ g\ Al} = 0.148\ mol\ available$

$$42.0\ g\ Br_2 \times \frac{1\ mol\ Br_2}{160\ g\ Br_2} = 0.263\ mol\ available$$

$$0.148\ mol\ Al \times \frac{3\ mol\ Br_2}{2\ mol\ Al} = 0.222\ mol\ Br_2\ needed$$

There is excess Br_2, so Al is the limiting reagent.

b. $0.148\ mol\ Al \times \frac{2\ mol\ AlBr_3}{2\ mol\ Al} \times \frac{267\ g\ AlBr_3}{1\ mol\ AlBr_3} = 39.5\ g\ AlBr_3$ theoretical yield

c. $\frac{32.2\ g\ actual\ yield}{39.5\ g\ theoretical\ yield} \times 100\% = 81.5\ \%\ yield$

11.23 a. $80.0\ kg\ C_2H_4 \times \frac{1\ kmol\ C_2H_4}{28.0\ kg\ C_2H_4} = 2.86\ kmol\ C_2H_4$ available

$$55.0\ kg\ H_2O \times \frac{1\ kmol\ H_2O}{18.0\ kg\ H_2O} = 3.06\ kmol\ H_2O\ available$$

$$2.86 \text{ kmol } C_2H_4 \times \frac{1 \text{ kmol } H_2O}{1 \text{ kmol } C_2H_4} = 2.86 \text{ kmol } H_2O \text{ needed}$$

C_2H_4 is the limiting reagent.

b. $$2.86 \text{ kmol } C_2H_4 \times \frac{1 \text{ kmol } C_2H_5OH}{1 \text{ kmol } C_2H_4} \times \frac{46.0 \text{ kg } C_2H_5OH}{1 \text{ kmol } C_2H_5OH} = 132 \text{ kg } C_2H_5OH \text{ theoretical yield}$$

c. $$\frac{125 \text{ kg actual yield}}{132 \text{ kg theoretical yield}} \times 100\% = 94.7 \text{ \% yield}$$

11.25

$$15.0 \text{ g Al} \times \frac{1 \text{ mol Al}}{27.0 \text{ g Al}} \times \frac{2 \text{ mol Fe}}{2 \text{ mol Al}} = 0.556 \text{ mol Fe}$$

$$30.0 \text{ g } Fe_2O_3 \times \frac{1 \text{ mol } Fe_2O_3}{159.6 \text{ g } Fe_2O_3} \times \frac{2 \text{ mol Fe}}{1 \text{ mol } Fe_2O_3} = 0.376 \text{ mol Fe}$$

Fe_2O_3 produces the smaller amount so it is the limiting reagent

$$0.376 \text{ mol Fe} \times \frac{55.8 \text{ g Fe}}{1 \text{ mol Fe}} = 20.9 \text{ g Fe}$$

11.27 During an exothermic reaction, heat is released; during an endothermic reaction, heat is absorbed.

11.29 $$10.0 \text{ g } CH_4 \times \frac{1 \text{ mol } CH_4}{16.0 \text{ g } CH_4} \times \frac{-802 \text{ kJ}}{1 \text{ mol } CH_4} = -501 \text{ kJ}$$

The enthalpy change is negative because the reaction is exothermic. The amount of energy released by burning methane is 3.5 times larger than that released by burning an equal number of grams of glucose.

11.31 $CaCO_3(s) \rightarrow CaO(s) + CO_2(g) \quad \Delta H = +178 \text{ kJ}$

$$1000. \text{ g CaO} \times \frac{1 \text{ mol CaO}}{56.0 \text{ g CaO}} \times \frac{+178 \text{ kJ}}{1 \text{ mol CaO}} = +3178 \text{ kJ}$$

11.33 a. increase b. increase

11.35

$$1000. \text{ miles} \times \frac{1 \text{ gal}}{36.2 \text{ mile}} \times \frac{4 \text{ qt}}{1 \text{ gal}} \times \frac{946 \text{ mL}}{1 \text{ qt}} \times \frac{0.8205 \text{ g } C_8H_{18}}{1.00 \text{ mL}} \times \frac{1 \text{ mol } C_8H_{18}}{114.0 \text{ g } C_8H_{18}} \times \frac{16 \text{ mol } CO_2}{2 \text{ mol } C_8H_{18}} \times$$

$$\frac{44.0 \text{ g } CO_2}{1 \text{ mol } CO_2} \times \frac{1 \text{ kg}}{1000 \text{ g}} = 265 \text{ kg } CO_2$$

11.37 a. $$0.0282 \text{ L } Na_2SO_4 \times \frac{0.100 \text{ mol } Na_2SO_4}{1.00 \text{ L}} \times \frac{1 \text{ mol } Pb^{2+}}{1 \text{ mol } Na_2SO_4} \times \frac{207.2 \text{ g } Pb^{2+}}{1 \text{ mol } Pb^{2+}} = 0.584 \text{ g } Pb^{2+} \text{ ions}$$

b. $0.584 \text{ g } Pb^{2+}/0.100 \text{ L} = 5.84 \text{ g/L}$

11.39 $$454. \text{ g } C_6H_{12}O_6 \times \frac{1 \text{ mol } C_6H_{12}O_6}{180. \text{ g } C_6H_{12}O_6} \times \frac{2 \text{ mol } C_2H_6O}{1 \text{ mol } C_6H_{12}O_6} \times \frac{46.0 \text{ g } C_2H_6O}{1 \text{ mol } C_2H_6O} = 232 \text{ g } C_2H_6O$$

11.41 a. $2\ C_4H_{10} + 13\ O_2(g) \rightarrow 8\ CO_2 + 10\ H_2O$

b. $5.00\text{ g } C_4H_{10} \times \frac{1\text{ mol } C_4H_{10}}{58.0\text{ g } C_4H_{10}} \times \frac{13\text{ mol } O_2}{2\text{ mol } C_4H_{10}} = 0.560\text{ mol } O_2$

c. $5.00\text{ g } C_4H_{10} \times \frac{1\text{ mol } C_4H_{10}}{58.0\text{ g } C_4H_{10}} \times \frac{13\text{ mol } O_2}{2\text{ mol } C_4H_{10}} \times \frac{32.0\text{ g } O_2}{1\text{ mol } O_2} = 17.9\text{ g } O_2$

11.43 $2\ Al_2O_3 \rightarrow 4\ Al + 3\ O_2$

$15.0\text{ ton } Al_2O_3 \times \frac{2000\text{ lb } Al_2O_3}{1\text{ ton } Al_2O_3} \times \frac{454\text{ g } Al_2O_3}{1\text{ lb } Al_2O_3} \times \frac{1\text{ mol } Al_2O_3}{102\text{ g } Al_2O_3} \times \frac{4\text{ mol Al}}{2\text{ mol } Al_2O_3} \times \frac{27.0\text{ g Al}}{1\text{ mol Al}} \times$

$\frac{1\text{ lb Al}}{454\text{ g Al}} \times \frac{1\text{ ton Al}}{2000\text{ lb Al}} = 7.94\text{ tons Al}$

11.45

$10.0\text{ ton } Fe_2O_3 \times \frac{2000\text{ lb } Fe_2O_3}{1\text{ ton } Fe_2O_3} \times \frac{454\text{ g } Fe_2O_3}{1\text{ lb } Fe_2O_3} \times \frac{1\text{ mol } Fe_2O_3}{159.7\text{ g } Fe_2O_3} \times \frac{2\text{ mol Fe}}{1\text{ mol } Fe_2O_3} \times \frac{55.8\text{ g Fe}}{1\text{ mol Fe}} \times$

$\frac{1\text{ lb Fe}}{454\text{ g Fe}} \times \frac{1\text{ ton Fe}}{2000\text{ lb Fe}} = 6.99\text{ ton Fe}$

11.47 $75.0\text{ g } Na_2O_2 \times \frac{1\text{ mol } Na_2O_2}{78.0\text{ g } Na_2O_2} \times \frac{2\text{ mol } CO_2}{2\text{ mol } Na_2O_2} \times \frac{44.0\text{ g } CO_2}{1\text{ mol } CO_2} = 42.3\text{ g } CO_2$

11.49 $0.0100\text{ L acid} \times \frac{2.00\text{ mol HCl}}{1\text{ L acid}} \times \frac{1\text{ mol } NaHCO_3}{1\text{ mol HCl}} \times \frac{84.0\text{ g } NaHCO_3}{1\text{ mol } NaHCO_3} = 1.68\text{ g } NaHCO_3$

11.51 a. $\frac{100\text{ cars}}{1\text{ day}} \times \frac{100\text{ tons coal}}{1\text{ car}} \times \frac{2\text{ tons } S_8}{100\text{ tons coal}} = 200.\text{ tons } S_8\text{ / day}$

b. $\frac{200\text{ g } S_8}{1\text{ day}} \times \frac{1\text{ mol } S_8}{256\text{ g } S_8} \times \frac{8\text{ mol } SO_2}{1\text{ mol } S_8} \times \frac{64.0\text{ g } SO_2}{1\text{ mol } SO_2} = 400.\text{ tons } SO_2$

We can work the problem using quantities in grams and substitute pounds in the answer; using grams or pounds throughout gives the same number.

11.53 a. $30.0\text{ g } NaNO_3 \times \frac{1\text{ mol } NaNO_3}{85.0\text{ g } NaNO_3} = 0.353\text{ mol } NaNO_3$ available

$0.022\text{ L acid} \times \frac{18\text{ mol } H_2SO_4}{1\text{ L acid}} = 0.396\text{ mol } H_2SO_4$ available

$0.353\text{ mol } NaNO_3 \times \frac{1\text{ mol } H_2SO_4}{1\text{ mol } NaNO_3} = 0.353\text{ mol } H_2SO_4$ needed

$NaNO_3$ is the limiting reagent.

$0.353\text{ mol } NaNO_3 \times \frac{1\text{ mol } HNO_3}{1\text{ mol } NaNO_3} \times \frac{63.0\text{ g } HNO_3}{1\text{ mol } HNO_3} = 22.2\text{ g } HNO_3$ theoretical yield

b. $\frac{17.0\text{ g actual yield}}{22.2\text{ g theoretical yield}} \times 100\% = 76.6\ \%$ yield

11.55 $1000.\ g\ O_3 \times \frac{1\ mol\ O_3}{48.00\ g\ O_3} \times \frac{3\ mol\ O_2}{2\ mol\ O_3} \times \frac{32.00\ g\ O_2}{1\ mol\ O_2} = 1000.\ g\ O_2$

11.57 $50.0\ kg\ NaOH \times \frac{1\ mol\ NaOH}{40.0\ g\ NaOH} \times \frac{1\ mol\ Cl_2}{2\ mol\ NaOH} \times \frac{71.0\ g\ Cl_2}{1\ mol\ Cl_2} = 44.4\ kg\ Cl_2$

11.59 $1000.\ L\ acid \times \frac{0.500\ mol\ HCl}{1\ L\ acid} \times \frac{1\ mol\ Ca(OH)_2}{2\ mol\ HCl} \times \frac{74.0\ g\ Ca(OH)_2}{1\ mol\ Ca(OH)_2} \times \frac{1\ kg}{1000\ g} = 18.5\ kg\ Ca(OH)_2$

11.61 a. $0.250\ L\ HCl(aq) \times \frac{2.00\ mol\ HCl}{1\ L\ HCl} \times \frac{1\ mol\ Cl_2}{1\ mol\ HCl} = 0.500\ mol\ Cl_2$

b. $0.250\ L\ HCl(aq) \times \frac{2.00\ mol\ HCl}{1\ L\ HCl} \times \frac{1\ mol\ Cl_2}{1\ mol\ HCl} \times \frac{71.0\ g\ Cl_2}{1\ mol\ Cl_2} = 35.5\ g\ Cl_2$

c. The reaction of bleach with acid produces Cl_2 gas which can be quite harmful if inhaled. The gas should not be released in a closed room.

11.63 $500.\ g\ Al \times \frac{1\ mol\ Al}{27.0\ g\ Al} \times \frac{+3340\ kJ}{4\ mol\ Al} = +1.55 \times 10^4\ kJ$

11.65 $0.500\ g\ CaCO_3 \times \frac{1\ mol\ CaCO_3}{100.1\ g\ CaCO_3} \times \frac{2\ mol\ HCl}{1\ mol\ CaCO_3} \times \frac{1.00\ L\ acid}{0.100\ mol\ HCl} \times \frac{1000\ mL}{1.00\ L} = 100.\ mL\ acid$

11.67 $8.64\ kg\ CaCO_3 \times \frac{1\ mol\ CaCO_3}{100.1\ g\ CaCO_3} \times \frac{1\ mol\ CaO}{1\ mol\ CaCO_3} \times \frac{56.0\ g\ CaO}{1\ mol\ CaO} = 4.84\ kg\ CaO$

11.69 a. $12.0\ mol\ NH_3 \times \frac{5\ mol\ O_2}{4\ mol\ NH_3} = 15.0\ mol\ O_2$

b. 14.0 mol oxygen; 15.0 mol of oxygen are needed to react with 12.0 mol ammonia.

c. $12.0\ mol\ NH_3 \times \frac{4\ mol\ NO}{4\ mol\ NH_3} = 12.0\ mol\ NO$

11.71 $\frac{12.8\ tons}{13.3\ tons} \times 100 = 96.2\%$

11.73 $4.00\ g\ Si \times \frac{1\ mol\ Si}{28.1\ g\ Si} \times \frac{1\ mol\ Si_3N_4}{3\ mol\ Si} = 0.0474\ mol\ Si_3N_4$

$3.00\ g\ N_2 \times \frac{1\ mol\ N_2}{28.0\ g\ N_2} \times \frac{1\ mol\ Si_3N_4}{2\ mol\ N_2} = 0.0536\ mol\ Si_3N_4$

The Si is the limiting reagent. $0.0474\ mol\ Si_3N_4 \times \frac{140.3\ g\ Si_3N_4}{1\ mol\ Si_3N_4} = 6.64\ g\ Si_3N_4$

11.75 $25.0\ g\ Fe \times \frac{1\ mol\ Fe}{55.8\ g\ Fe} \times \frac{2\ mol\ Fe_2O_3}{4\ mol\ Fe} \times \frac{159.7\ g\ Fe_2O_3}{1\ mol\ Fe_2O_3} = 35.8\ g\ Fe_2O_3$

35.8 g – 25.0 g = 10.8 g increase $\frac{10.8\ g}{25.0\ g} x 100 = 43.2\%$

11.77 $Zn(s) + 2HCl(aq) \rightarrow ZnCl_2(aq) + H_2(g)$

$$28.7\ \text{mL} \times \frac{1\ \text{L}}{1000\ \text{mL}} \times \frac{0.250\ \text{mol HCl}}{1\ \text{L}} \times \frac{1\ \text{mol Zn}}{2\ \text{mol HCl}} \times \frac{65.4\ \text{g Zn}}{1\ \text{mol Zn}} = 0.235\ \text{g Zn}$$

11.79 $$12.0\ \text{g}\ C_6H_{12}O_6 \times \frac{1\ \text{mol}\ C_6H_{12}O_6}{180.0\ \text{g}\ C_6H_{12}O_6} \times \frac{6\ \text{mol}\ CO_2}{1\ \text{mol}\ C_6H_{12}O_6} \times \frac{44.0\ \text{g}\ CO_2}{1\ \text{mol}\ CO_2} = 17.6\ \text{g}\ CO_2$$

11.81 $2\ C_{14}H_{30} + 43\ O_2 \rightarrow 28\ CO_2 + 30\ H_2O$

$$12.0\ \text{kg}\ C_{14}H_{30} \times \frac{1\ \text{kmol}\ C_{14}H_{30}}{198.0\ \text{kg}\ C_{14}H_{30}} \times \frac{43\ \text{kmol}\ O_2}{2\ \text{kmol}\ C_{14}H_{30}} \times \frac{32.0\ \text{kg}\ O_2}{1\ \text{kmol}\ O_2} = 41.7\ \text{kg}\ O_2$$

11.83 $Al_2(SO_4)_3 + 3\ BaCl_2 \rightarrow 2\ AlCl_3 + 3\ BaSO_4$

$$15.0\ \text{mL} \times \frac{1\ \text{L}}{1000\ \text{mL}} \times \frac{0.100\ \text{mol}\ Al_2(SO_4)_3}{1\ \text{L}} \times \frac{3\ \text{mol}\ BaCl_2}{1\ \text{mol}\ Al_2(SO_4)_3} \times \frac{1\ \text{L}}{0.100\ \text{mol}\ BaCl_2} \times \frac{1000\ \text{mL}}{1\ \text{L}} = 45.0\ \text{mL}$$

11.85

$$10.0\ \text{ton PbS} \times \frac{2000\ \text{lb PbS}}{1\ \text{ton PbS}} \times \frac{454\ \text{g PbS}}{1\ \text{lb PbS}} \times \frac{1\ \text{mol PbS}}{239.0\ \text{g PbS}} \times \frac{2\ \text{mol PbO}}{2\ \text{mol PbS}} \times \frac{223.2\ \text{g PbO}}{1\ \text{mol PbO}} \times$$

$$\frac{1\ \text{lb PbO}}{454\ \text{g PbO}} \times \frac{1\ \text{ton PbO}}{2000\ \text{lb PbO}} = 9.33\ \text{ton PbO}$$

11.87

$$20.0\ \text{ton ZnS} \times \frac{2000\ \text{lb ZnS}}{1\ \text{ton ZnS}} \times \frac{454\ \text{g ZnS}}{1\ \text{lb ZnS}} \times \frac{1\ \text{mol ZnS}}{97.5\ \text{g ZnS}} \times \frac{2\ \text{mol ZnO}}{2\ \text{mol ZnS}} \times \frac{81.4\ \text{g ZnO}}{1\ \text{mol ZnO}} \times$$

$$\frac{1\ \text{lb ZnO}}{454\ \text{g ZnO}} \times \frac{1\ \text{ton ZnO}}{2000\ \text{lb ZnO}} = 16.7\ \text{ton ZnO}$$

CHAPTER 12
Gases

12.1 Gas particles are moving rapidly, randomly, and continuously in all directions with negligible attraction between them.

12.3 If the volume of the container is decreased, the particles will travel shorter distances before they strike the walls. So they strike the walls more frequently, thus increasing the pressure -- the force per unit area.

12.5 When the speed of the particles decreases, the temperature also decreases.

12.7 The two gases have the same kinetic energy when they are at the same temperature, but the He atoms move faster. The tank of He is lighter but the number of particles in both tanks is the same because equal volumes contain equal numbers of particles at the same conditions.

12.9 Oxygen sample A which is at a higher pressure also has a higher density than oxygen sample B because there are more molecules per unit volume for sample A.

12.11 Because gas molecules move continuously, rapidly and randomly in all directions, the aroma moves throughout all available space.

12.13 Additional air can be added to a pressurized tire because the particles of air are extremely tiny and the distance between them is large.

12.15 Earth's atmosphere is the layer of gases that surrounds our planet.

12.17 Pressure is the force exerted per unit area. Atmospheric pressure is the total force exerted by the air on each unit of area.

12.19 The mercury in a mercury barometer flows in or out to maintain equal pressure on the inside and outside of the tube. A measurement of the height of the mercury column is reported as the barometric pressure.

12.21 a. $1.00 \text{ atm} \times \frac{760 \text{ torr}}{1 \text{ atm}} = 760. \text{ torr}$

b. $912 \text{ torr} \times \frac{1 \text{ atm}}{760 \text{ torr}} = 1.20 \text{ atm}$

c. $0.500 \text{ atm} \times \frac{760 \text{ mm Hg}}{1 \text{ atm}} = 380. \text{ mmHg}$

d. $1200 \text{ psi} \times \frac{1 \text{ atm}}{14.7 \text{ psi}} = 81.6 \text{ atm}$

e. $2.00 \text{ atm} \times \frac{101 \text{ kPa}}{1 \text{ atm}} = 202 \text{ kPa}$

f. $3.0 \text{ in. Hg} \times \frac{1 \text{ atm}}{29.9 \text{ in. Hg}} = 0.10 \text{ atm}$

12.23 Pressure times volume equals a constant for a sample of gas at a fixed temperature. Thus, as pressure increases, volume decreases, as shown by the bicycle tire pump model: when the volume of trapped gas decreases - by pressing in on the plunger - the pressure of the gas increases.

12.25 7.45×10^4 for each set of data.

12.27 $P_1V_1 = P_2V_2$

$$P_2 = \frac{P_1V_1}{V_2} = \frac{1.00 \text{ atm x } 400.\text{ cm}^3}{100.\text{ cm}^3} = 4.00 \text{ atm}$$

12.29 $P_1V_1 = P_2V_2$

$$V_2 = \frac{P_1V_1}{P_2} = \frac{2200.\text{ psi x } 60.0 \text{ L}}{14.7 \text{ psi}} = 8.98 \times 10^3 \text{ L} \quad 8.98 \times 10^3 \text{ L x } \frac{\text{min}}{8.00 \text{ L}} = 1120 \text{ min or } 18.7 \text{ hours}$$

12.31 At constant pressure, volume increases as temperature increases. Volume is proportional to temperature. The graph of V vs. T is a straight line.

12.33 4.00 L/373 K = 0.0107, 2.93 L/272 K = 0.0107, etc. The V/T ratios for this example equal 0.0107 using Kelvin temperatures. V/T ratios are constant only with Kelvin temperatures.

12.35 $V_1 = 400.$ mL $\quad V_2 = ?$ L $\quad T_1 = -120$ °C (153. K) $\quad T_2 = 100.$ °C (373 K)

$$\frac{V_1}{T_1} = \frac{V_2}{T_2} \qquad V_2 = \frac{V_1T_2}{T_1} = \frac{400.\text{ mL x } 373 \text{ K}}{153 \text{ K}} = 975 \text{ mL}$$

12.37 $V_1 = 1500.$ mL $\quad V_2 = 750.$ mL $\quad T_1 = 22$ °C (295 K) $\quad T_2 = ?$

$$\frac{V_1}{T_1} = \frac{V_2}{T_2} \qquad T_2 = \frac{T_1V_2}{V_1} = \frac{295 \text{ K x } 750.\text{ mL}}{1500.\text{ mL}} = 148 \text{ K or } -125\ ^\circ\text{C} \qquad \text{a change of } 147\ ^\circ\text{C}$$

12.39 As T increases at constant V, the average kinetic energy of the particles increases and gas particles move faster. The fast-moving particles bombard the walls of the container more frequently, so pressure increases.

12.41 $P_1 = 720.$ torr $\quad P_2 = ?$ torr $\quad T_1 = 20$ °C (293 K) $\quad T_2 = 750$ °C (1023 K)

$$\frac{P_1}{T_1} = \frac{P_2}{T_2} \qquad P_2 = \frac{P_1T_2}{T_1} = \frac{720.\text{ torr x } 1023 \text{ K}}{293.\text{ K}} = 2.51 \times 10^3 \text{ torr}$$

12.43 $P_1 = 30. + 14.7$ psi $\quad P_2 = 34 + 14.7$ psi $\quad T_1 = 20.$ °C (293 K) $\quad T_2 = ?$

$$\frac{P_1}{T_1} = \frac{P_2}{T_2} \qquad T_2 = \frac{T_1P_2}{P_1} = \frac{293 \text{ K x } 48.7 \text{ psi}}{44.7 \text{ psi}} = 319 \text{ K or } 46\ ^\circ\text{C}$$

12.45 $P_1 = 1.96$ atm $\quad V_1 = 45.2$ mL $\quad T_1 = 21$ °C (294 K)
$P_2 = 1.00$ atm $\quad V_2 = ?$ L $\quad T_2 = 0$ °C (273 K)

$$\frac{P_1V_1}{T_1} = \frac{P_2V_2}{T_2} \qquad V_2 = \frac{V_1P_1T_2}{T_1P_2} = \frac{45.2 \text{ mL x } 1.96 \text{ atm x } 273 \text{ K}}{294 \text{ K x } 1.00 \text{ atm}} = 82.3 \text{ mL}$$

12.47 $P_1 = 1.00$ atm $\quad V_1 = 800.$ mL $\quad T_1 = 10.$ °C (283 K)
$P_2 = ?$ $\quad V_2 = 850.$ mL $\quad T_2 = 100$ °C (373 K)

$$\frac{P_1V_1}{T_1} = \frac{P_2V_2}{T_2} \qquad P_2 = \frac{P_1V_1T_2}{T_1V_2} = \frac{1.00 \text{ atm x } 800.\text{ mL x } 373 \text{ K}}{850.\text{ mL x } 283 \text{ K}} = 1.24 \text{ atm}$$

12.49 $P_1 = 1.00$ atm $\quad V_1 = 22.4$ L $\quad T_1 = 0$ °C (273 K) $\quad$ at STP conditions
$P_2 = 1.20$ atm $\quad V_2 = ?$ L $\quad T_2 = 100$ °C (373 K)

$$\frac{P_1V_1}{T_1} = \frac{P_2V_2}{T_2} \qquad V_2 = \frac{V_1P_1T_2}{T_1P_2} = \frac{22.4 \text{ L x } 1.00 \text{ atm x } 373 \text{ K}}{273 \text{ K x } 1.20 \text{ atm}} = 25.5 \text{ L}$$

Percent increase in volume: $\frac{\text{Volume increase}}{\text{Original volume}} \text{ x } 100\% = \frac{3.1 \text{ L}}{22.4 \text{ L}} \text{ x } 100\% = 13.8\ \%$

12.51 $\frac{V_1}{n_1} = \frac{V_2}{n_2}$ The volume of gas at constant temperature and pressure is proportional to the number of moles (n) of a gas.

12.53 Since the volume is double at the same temperature and pressure the number of molecules will be double.

12.55 $\frac{V_1}{n_1} = \frac{V_2}{n_2} \quad V_2 = \frac{V_1n_2}{n_1} = \frac{6.38 \text{ L x } 0.450 \text{ mol}}{0.250 \text{ mol}} = 11.5 \text{ L}$

12.57 Start with the mass given and use the molar mass of O_2 (g/mol) and molar volume at STP to obtain the volume of gas at STP.

$$0.200 \text{ g } O_2 \text{ x } \frac{1 \text{ mol}}{32.0 \text{ g } O_2} \text{ x } \frac{22.4 \text{ L}}{1 \text{ mol}} = 0.140 \text{ L}$$

12.59 Start with the volume given and use the molar volume at STP and the molar mass of CO_2 (g/mol) to obtain the mass of CO_2 for the sample.

$$4.00 \text{ L } CO_2 \text{ x } \frac{1 \text{ mol}}{22.4 \text{ L } CO_2} \text{ x } \frac{44.0 \text{ g}}{1 \text{ mol}} \text{ x } \frac{1000 \text{ mg}}{1 \text{ g}} = 7.86 \text{ x } 10^3 \text{ mg}$$

12.61 Start with the volume of sample given and use the molar volume at STP to obtain the number of moles for the gas sample.

$$6.00 \text{ L x } \frac{1 \text{ mol}}{22.4 \text{ L}} = 0.268 \text{ mol}$$

12.63 To obtain the density of gas at STP in g/L, start with the molar mass in g/mol and use the molar volume as a conversion factor.

$$\frac{28.0 \text{ g } N_2}{1 \text{ mol } N_2} \text{ x } \frac{1 \text{ mol}}{22.4 \text{ L}} = 1.25 \text{ g / L}$$

12.65 To obtain the density of gas at STP in g/L, start with the molar mass in g/mol and use the molar volume as a conversion factor.

$$\frac{71.0 \text{ g}}{1 \text{ mol } Cl_2} \text{ x } \frac{1 \text{ mol}}{22.4 \text{ L}} = 3.17 \text{ g / L}$$

12.67 An ideal gas is one that perfectly obeys the gas laws. There is no ideal gas; only "real" gases actually exist.

12.69 $\frac{\text{Liters x atmospheres}}{\text{moles x kelvins}}$

12.71 $P = \frac{nRT}{V} = \frac{44.0 \text{ mol x } (0.0821 \text{ L} - \text{atm/mol} - \text{K}) \text{ x } 295 \text{ K}}{36 \text{ L}} = 30. \text{ atm}$

12.73 First convert pressure to atm: 800. torr x 1 atm/760 torr = 1.05 atm

$$g = \frac{M_m PV}{RT} = \frac{4.00 \text{ g / mol x } 1.05 \text{ atm x } 8.50 \text{ L}}{(0.0821 \text{ L} - \text{atm / mol} - \text{K}) \text{ x } 293 \text{ K}} = 1.48 \text{ g}$$

12.75 Total pressure = 0.25 atm + 0.50 atm + 0.20 atm = 0.95 atm

12.77 100. atm total pressure x 0.970 = 97.0 atm of CO_2 on Venus.

12.79 $P_{oxygen} = P_{total} - P_{water\ vapor} = 740.\ \text{torr} - 23.8\ \text{torr} = 716\ \text{torr}$

$$n = \frac{PV}{RT} = \frac{(716 / 760\ \text{atm}) \times 0.0950\ \text{L}}{(0.0821\ \text{L} - \text{atm} / \text{mol} - \text{K}) \times 298\ \text{K}} = 3.66 \times 10^{-3}\ \text{mol}$$

12.81 a. $1\ \text{tank acetylene} \times \frac{5\ \text{mol}\ O_2}{2\ \text{mol acetylene}} = 2.50\ \text{tanks}\ O_2$

b. The volume ratio of oxygen to acetylene is 5 to 2 when pressures are the same for both gases, so the tank of oxygen needs to be 2.5 times as large as the acetylene tank if the tanks are to become empty at the same time.

12.83 $2\ CO(g) + O_2(g) \rightarrow 2\ CO_2(g)$ $\quad 38.0\ \text{L CO} \times \frac{2\ \text{L}\ CO_2}{2\ \text{L CO}} = 38.0\ \text{L}\ CO_2$

12.85 $160.\ \text{L} \times \frac{5\ \text{L}\ O_2}{4\ \text{L}\ NH_3} = 200.\ \text{L}\ O_2$

12.87 $C_6H_{12}O_6 \rightarrow 2\ C_2H_5OH + 2\ CO_2$

Convert grams of glucose to moles of CO_2. Then use PV = nRT to find V.

$$500.\ \text{g glucose} \times \frac{1\ \text{mol glucose}}{180.\ \text{g glucose}} \times \frac{2\ \text{mol}\ CO_2}{1\ \text{mol glucose}} = 5.56\ \text{mol}\ CO_2$$

$$V = \frac{nRT}{P} = \frac{5.56\ \text{mol} \times (0.0821\ \text{L} - \text{atm} / \text{mol} - \text{K}) \times 293\ \text{K}}{1\ \text{atm}} = 134\ \text{L}$$

12.89 Determine the number of moles of H_2 produced by 2.15 g of Mg.

$$2.15\ \text{g Mg} \times \frac{1\ \text{mol Mg}}{24.3\ \text{g Mg}} \times \frac{1\ \text{mol}\ H_2}{1\ \text{mol Mg}} = 0.0855\ \text{mol}\ H_2$$

At STP the volume of H_2 gas produced is

$$0.0855\ \text{mol}\ H_2 \times \frac{22.4\ \text{L}}{1\ \text{mol}} = 1.98\ \text{L}\ H_2\ \text{at STP}$$

At 735 torr and 25 °C, the volume of H_2 gas produced is

$$V = \frac{nRT}{P} = \frac{0.0855\ \text{mol} \times (0.0821\ \text{L} - \text{atm} / \text{mol} - \text{K}) \times 298\ \text{K}}{735 / 760\ \text{atm}} = 2.24\ \text{L at 735 torr}$$

12.91 $6\ NaN_3 + Fe_2O_3 \rightarrow 3\ Na_2O + 2\ Fe + 9\ N_2$

Determine the moles of N_2 in the air bag.

$$n = \frac{PV}{RT} = \frac{(1.20\ \text{atm})(5.50\ \text{L})}{(0.0821\ \text{L} - \text{atm/mol} - \text{K})(291\ \text{K})} = 0.276\ \text{mol}\ N_2$$

$$0.276\ \text{mol}\ N_2 \times \frac{6\ \text{mol}\ NaN_3}{9\ \text{mol}\ N_2} \times \frac{65.0\ \text{g}\ NaN_3}{1\ \text{mol}\ NaN_3} = 12.0\ \text{g}\ NaN_3$$

12.93 $CaCO_3 + 2\ HCl \rightarrow CaCl_2 + H_2O + CO_2(g)$

First determine the number of moles of CO_2 that could be produced from each reagent.

$$0.500\ \text{g}\ CaCO_3 \times \frac{1\ \text{mol}\ CaCO_3}{100.\ \text{g}\ CaCO_3} \times \frac{1\ \text{mol}\ CO_2}{1\ \text{mol}\ CaCO_3} = 0.00500\ \text{mol}\ CO_2$$

$$0.0750 \text{ L HCl} \times \frac{0.125 \text{ mol HCl}}{1.00 \text{ L}} \times \frac{1 \text{ mol } CO_2}{2 \text{ mol HCl}} = 0.00469 \text{ mol } CO_2$$

The 0.0750 L of HCl solution is the limiting reagent so 0.00469 mol of CO_2 will be produced.

$$V = \frac{nRT}{P} = \frac{0.00469 \text{ mol} \times (0.0821 \text{ L} - \text{atm / mol} - \text{K}) \times 295 \text{ K}}{742 / 760 \text{ atm}} = 0.116 \text{ L } CO_2$$

12.95 $2\, C_8H_{18} + 25\, O_2 \rightarrow 16\, CO_2 + 18\, H_2O$

$$1.00 \text{ gal} \times \frac{4 \text{ qt}}{1 \text{ gal}} \times \frac{946 \text{ mL}}{1 \text{ qt}} \times \frac{0.807 \text{ g } C_8H_{18}}{1.00 \text{ mL}} \times \frac{1 \text{ mol } C_8H_{18}}{114 \text{ g } C_8H_{18}} \times \frac{16 \text{ mol } CO_2}{2 \text{ mol } C_8H_{18}} = 214 \text{ mol } CO_2$$

$$V = \frac{nRT}{P} = \frac{214 \text{ mol} \times (0.0821 \text{ L} - \text{atm / mol} - \text{K}) \times 292 \text{ K}}{735 / 760 \text{ atm}} = 53\bar{0}0 \text{ L } CO_2$$

12.97 A bicycle tire pump.

12.99 To obtain gas density at STP in g/L, start with the molar mass in g/mol and use the molar volume as a conversion factor.

$$\frac{17.0 \text{ g } NH_3}{1 \text{ mol } NH_3} \times \frac{1 \text{ mol}}{22.4 \text{ L}} = 0.759 \text{ g / L}$$

12.101 To obtain molar mass (g/mol) at STP, when the gas density is given, use the molar volume as a conversion factor.

$$\frac{1.96 \text{ g}}{1 \text{ L}} \times \frac{22.4 \text{ L}}{1 \text{ mol}} = 43.9 \text{ g / mol}$$

12.103 $V = \frac{nRT}{P} = \frac{0.600 \text{ mol} \times (0.0821 \text{ L atm / mol K}) \times 310 \text{ K}}{0.800 \text{ atm}} = 19.1 \text{ L}$

12.105 $2\, C_8H_{18} + 25\, O_2 \rightarrow 16\, CO_2 + 18\, H_2O$

Determine the number of moles of H_2O produced by 500. g of C_8H_{18}.

$$500. \text{ g } C_8H_{18} \times \frac{1 \text{ mol } C_8H_{18}}{114 \text{ g } C_8H_{18}} \times \frac{18 \text{ mol } H_2O}{2 \text{ mol } C_8H_{18}} = 39.5 \text{ mol } H_2O$$

At standard pressure (1 atm) and 18 °C, the volume of H_2O gas produced is

$$V = \frac{nRT}{P} = \frac{39.5 \text{ mol} \times (0.0821 \text{ L atm / mol K}) \times 291 \text{ K}}{1 \text{ atm}} = 944 \text{ L } H_2O \text{ vapor}$$

12.107 $\frac{16.0 \text{ g}}{1 \text{ mol } CH_4} \times \frac{1 \text{ mol}}{22.4 \text{ L}} = 0.714 \text{ g / L}$

12.109 $V = \frac{nRT}{P} = \frac{1 \text{ mol} \times (0.0821 \text{Latm/molK}) \times 295 \text{ K}}{0.984 \text{atm}} = 24.6 \text{ L } C_3H_8$ $\quad \frac{44.0 \text{ g}}{24.6 \text{ L}} = 1.79 \text{ g/L}$

12.111 0.80 x 784 torr = 630 torr N_2 $\quad$ 0.20 x 784 torr = 160 torr O_2

12.113 $\frac{P_1}{T_1} = \frac{P_2}{T_2}$ $\quad \frac{235 \text{ atm}}{295 \text{ K}} = \frac{P_2}{77 \text{ K}}$ $\quad P_2 = 61.3 \text{ atm}$

12.115 $C_3H_8 + 5\, O_2 \rightarrow 3\, CO_2 + 4\, H_2O$ $\quad 16.0 \text{ L} \times \frac{5 \text{ L } O_2}{1 \text{ L } C_3H_8} = 80.0 \text{ L } O_2$

12.117 $25.0 \text{ g Fe} \times \frac{1 \text{ mol Fe}}{55.8 \text{ g Fe}} \times \frac{3 \text{ mol } O_2}{4 \text{ mol Fe}} = 0.336 \text{ mol } O_2$

$$V = \frac{nRT}{P} = \frac{0.336 \text{ mol x } (0.0821 \text{ Latm/molK}) \text{ x } 292\text{K}}{0.978 \text{ atm}} = 8.24 \text{ L } O_2$$

12.119 $12.0 \text{ ml x } \frac{1 \text{ L}}{1000 \text{ mL}} \text{ x } \frac{0.840 \text{ mol } CH_3COOH}{1 \text{ L}} \text{ x } \frac{1 \text{ mol } CO_2}{1 \text{ mol } CH_3COOH} = 0.0101 \text{ mol } CO_2$

$$0.0101 \text{ mol } CO_2 \text{ x } \frac{22.4 \text{ L}}{1 \text{ mol}} = 0.226 \text{ L } CO_2$$

CHAPTER 13
Liquids and Solids

13.1 In both liquids and solids, particles are in contact with one another so liquids and solids have a specific volume. Particles of a solid are fixed in a rigid structure, but particles in liquids are free to move about in a limited extent as long as they remain in contact.

13.3 $$24.0 \text{ L water} \times \frac{1000 \text{ mL}}{1 \text{ L}} \times \frac{1.00 \text{ g water}}{1 \text{ mL water}} \times \frac{1 \text{ mol water}}{18.0 \text{ g water}} \times \frac{6.02 \times 10^{23} \text{ molecules}}{1 \text{ mol water}} = 8.03 \times 10^{23} \text{ molecules}$$

13.5 The halogens with the largest molar masses are more likely to be in the solid state than in the liquid or gaseous state.

13.7 a. nonpolar covalent

b. ionic

13.9 a. This solid is a metal (it conducts electricity)

b. This solid (perhaps table sugar) has polar covalent molecules.

13.11 *Inter*molecular forces are the attractions of molecules for one another. *Intra*molecular forces are those that exist within a molecule due to bonding.

13.13 Four types of intermolecular forces plus interionic forces, in order of increasing strength: dispersion forces, dipole forces, hydrogen bonding, interionic forces.

13.15 a. London forces

b. hydrogen bonding, dipole, and London forces

c. dipole forces and London forces

d. dipole forces and London forces

e. hydrogen bonding, dipole, and London forces

13.17 I_2 (London forces) $<$ HI (London and dipole forces) $<$ NH_3 (hydrogen bonding plus London and dipole forces) $<$ $NaNO_3$ (interionic forces).

13.19 London forces are greater in Br_2 (larger molecules) than in F_2 molecules.

13.21 Both CCl_4 and CF_4 are similar nonpolar molecules, but CCl_4 – with the greater molar mass and greater London forces – has a higher boiling point.

13.23 Only CH_3OH (with oxygen to hydrogen covalent bonds) has hydrogen bonding.

13.25 a. dipole and London forces

b. hydrogen bonding, dipole, and London forces

c. dipole and London forces

d. London forces

13.27 Liquids whose molecules have strong intermolecular forces also have strong intermolecular attractions, so they also have higher surface tension.

13.29 Water beads up on a freshly waxed surface—made up of nonpolar molecules—because the attraction of water molecules for one another is greater than is their attraction for the waxed surface. The "beading" does not occur on dirty surfaces covered with polar or ionic substances that dissolve in or are attracted to the polar water molecules.

13.31 Small, symmetrical molecules generally have low viscosities.

13.33 Viscosity increases at low temperatures, so low-viscosity motor oil is desirable in the winter when high-viscosity motor oil provides too much resistance to the movement of a piston in a cylinder. The low viscosity motor oil flows too readily (is too thin) at summer temperatures.

13.35 When water in a closed bottle is at equilibrium, liquid is evaporating at the same rate vapor is condensing; the mass of liquid remains constant.

13.37 The boiling point of a liquid is the temperature at which its vapor pressure becomes equal to the atmospheric pressure.

13.39 At high altitudes, atmospheric pressure is lower so the boiling point of water is lower, so food is cooked at a lower temperature. This increases the time required for complete cooking. When oil used for frying is below its boiling point, temperature and frying time are not affected.

13.41 B.P. low to high: O_2 (nonpolar) < CO (polar) < H_2O (hydrogen bonding).

13.43 Ethyl alcohol has the higher boiling point; its molecules undergo hydrogen bonding.

13.45 When a pot of potatoes is already boiling, turning up the heat will not cook the potatoes more quickly (the boiling temperature is not increased), it simply causes the water to boil away more quickly.

13.47 The atmospheric pressure is lower at higher elevations. The water boils at the temperature at which the vapor pressure is equal to the atmospheric pressure.

13.49 Explain to the student that the thermometer is not broken. The thermometer levels off at both 79 °C and 99 °C because these are the boiling points of the two fractions being seperated. Each fraction is distilled off before the temperature climbs further.

13.51 $$\frac{1368\ \text{J}}{1\ \text{g}} \times \frac{17.0\ \text{g}}{1\ \text{mol}} \times \frac{1\ \text{kJ}}{1000\ \text{J}} = 23.3\ \text{kJ/mol}$$

13.53 $$7.50\ \text{g} \times \frac{1\ \text{mol}}{58.0\ \text{g}} \times \frac{7.23\ \text{kcal}}{\text{mol}} = 0.935\ \text{kcal}$$

13.55 Carbon atoms in diamond are bonded to four other carbon atoms in three-dimensional tetrahedral arrangements. In graphite, hexagonal rings of carbon atoms join to form stacked planes (or layers) of carbon atoms. Both allotropes are crystalline, with carbon atoms joined covalently.

13.57 Both have six-membered rings of carbon atoms in hexagonal arrangements. In graphite these arrangements are in flat sheets, but in nanotubes the carbon rings form cylindrical, thread-like strands that are extremely strong.

13.59 In both S_8 and graphite, atoms form covalently bonded rings.

13.61 The properties of aspirin are like those of molecular solids.

13.63 Energy is absorbed by a substance during melting. Energy is released by a substance during freezing. These are reverse processes. The energy absorbed during melting is the same as that released during freezing.

13.65 $$15.0\text{ kg} \times \frac{1000.\text{ g}}{\text{kg}} \times \frac{80.\text{ cal}}{\text{g}} \times \frac{1\text{ kcal}}{1000\text{ cal}} = 1200\text{ kcal}$$

13.67 To warm the ice: $$75.0\text{ g} \times 5\ ^\circ\text{C} \times \frac{2.09\text{ J}}{\text{g }^\circ\text{C}} \times \frac{1\text{ kJ}}{1000\text{ J}} = 0.784\text{ kJ}$$

To melt the ice: $$75.0\text{ g} \times \frac{1\text{ mol}}{18.0\text{ g}} \times \frac{5.98\text{ kJ}}{\text{mol}} = 24.9\text{ kJ}$$

To warm the water: $$75.0\text{ g} \times 100\ ^\circ\text{C} \times \frac{4.18\text{ J}}{\text{g }^\circ\text{C}} \times \frac{1\text{ kJ}}{1000\text{ J}} = 31.4\text{ kJ}$$

To vaporize the water: $$75.0\text{ g} \times \frac{1\text{ mol}}{18.0\text{ g}} \times \frac{40.7\text{ kJ}}{\text{mol}} = 170.\text{ kJ}$$

The total energy change is 0.784 + 24.9 kJ + 31.4 kJ + 170. kJ = 227 kJ absorbed

13.69 To melt the snow: Table 13.5 gives the heat of fusion of water as 1.44 kcal /mol

$$5.00\text{ g} \times \frac{1\text{ mol}}{18.0\text{ g}} \times \frac{1.44\text{ kcal}}{\text{mol}} = 0.400\text{ kcal}$$

To warm the water: $$5.00\text{ g} \times 40.\ ^\circ\text{C} \times \frac{1.00\text{ cal}}{\text{g }^\circ\text{C}} \times \frac{1\text{ kcal}}{1000\text{ cal}} = 0.200\text{ kcal}$$

The total energy change is 0.400 kcal + 0.200 kcal = 0.600 kcal or 600 cal

13.71 Unlike nearly every other substance, water is less dense when it is solid (ice) than when it is liquid. That is why ice floats in liquid water.

13.73 The specific heat of water, 1 cal/g-°C, is higher than any other substance listed in Table 13.6. Thus, great amounts of heat are needed to warm a given quantity of water. However, the water has a great capacity to hold the heat, so vast amounts of water on the Earth alternately store and release heat. This helps moderate daily temperature fluctuations.

13.75 When water dissolves ionic compounds, the ion to ion attractions are overcome by many water molecules surrounding each ion in solution.

13.77 Water will hydrogen bond to itself requiring more energy to separate the molecules. Gasoline is a non-polar molecule having only weak London forces.

13.79 Ammonia hydrogen bonds, has an intermediate melting point and has fragile crystals. Diamonds are covalently bonded, does not melt and is very hard.

13.81 Molecules in liquid O_2 are held by London forces of attraction.

13.83 Ethylene glycol has two -OH groups that can enter into hydrogen bonding so it would be expected to have a higher boiling point than ethanol which has only one -OH group. Ethylene glycol could also be expected to evaporate more slowly than ethanol for the same reason.

13.85 Because oil evaporates more slowly than gasoline, intermolecular forces in the oil must be greater. (This is because molecules in the oil are larger – and have a greater molar mass – than those in the gasoline.

13.87 Xenon can be expected to have a higher boiling point than neon because xenon has a greater molar mass—and greater London forces—than neon.

13.89 The suggestion is a good one. (The approach is actually used in sugar refining.) Under reduced pressure the water boils off at a much lower temperature so there is no danger of the sugar being burned.

13.91 The bonding in CO_2 is covalent (it's a gas) and the information given tells us it is nonpolar. It cannot have hydrogen bonding so its intermolecular forces are primarily London forces.

13.93 $500.\ \text{g} \times \frac{1\ \text{mol}}{121\ \text{g}} \times \frac{35.0\ \text{kJ}}{\text{mol}} = 145\ \text{kJ}$

13.95

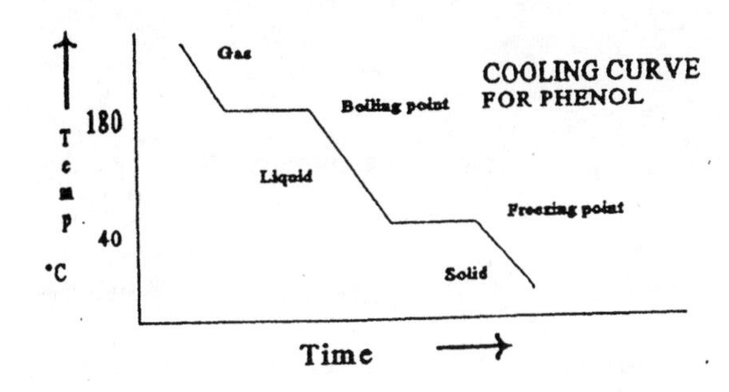

13.97 The heat of vaporization of water is 40.7 kJ/mol. That is the amount of heat energy difference between a mole of liquid water and a mole of steam at 100 °C. Burns caused by steam are more severe for this reason.

13.99 $50.0\ \text{g} \times \frac{1.37\ \text{kJ}}{\text{g}} = 68.5\ \text{kJ}$

13.101 Octadecane would be more viscous. Both chemicals are nonpolar (they contain only C and H atoms) and the larger molecule, which has greater London dispersion forces, would have greater viscosity.

13.103 Octane would evaporate more quickly. Both chemicals are nonpolar (they contain only C and H atoms) so the smaller molecule, which has weaker London dispersion forces and less attraction, would evaporate more quickly.

13.105 a. vaporization b. condensation

13.107 Both arrangements have an atom at the corners of a cube, but in the body-centered cubic arrangement, another atom is inside the cube at the center.

13.109 Alkali metals have body-centered cubic arrangements, but copper, silver and gold have face-centered cubic arrangements.

CHAPTER 14
Solutions

14.1 The solute is the substance being dissolved. The solvent dissolves the solute and is the substance present in the greater amount. A solution is a homogeneous mixture of solute and solvent.

14.3 a. Solute: carbon dioxide   Solvent: water   Solution state: liquid

b. Solute: nickel   Solvent: copper   Solution state: solid

c. Solute: ethanol   Solvent: water   Solution state: liquid

14.5 A small amount of table sugar or table salt is soluble in a glass of water, but large amounts of sugar or salt are not totally soluble in the water.

14.7 They are partially miscible.

14.9 In a saturated solution, the solute concentration is at a maximum. In a dilute solution, the solute concentration is less than the maximum.

14.11 Nonpolar solutes dissolve best in nonpolar solvents. Highly polar solutes dissolve best in polar solvents.

14.13 Glucose has several –OH groups that can be involved in hydrogen bonding with water, so the solubility of glucose is much greater than would be expected on the basis of molar mass.

14.15 a. Soluble; the compound is polar.

b. Soluble; the compound is polar and has a –OH group.

c. Soluble; nearly all compounds containing the ammonium ion are water soluble.

d. Soluble; all compounds containing the sodium ion are water soluble.

e. Insoluble; the compound is nonpolar.

f. Soluble; all compounds containing the potassium ion are water soluble.

14.17 a. Soluble; the Rb^+ is an alkali metal ion.

b. Soluble; all ammonium salts are soluble.

c. Insoluble; there are more than four carbon atoms per oxygen atom.

d. Soluble; the Na^+ is an alkali metal ion.

14.19 Ammonia is quite polar so it does not dissolve in hexane (which is nonpolar), but it readily dissolves in water due to hydrogen bonding.

14.21 When ionic solids dissolve in water, the energy released by the interaction of the solute with water must be greater than the sum of the energy needed to overcome the forces holding the ions together in the crystal and the energy needed to separate the water molecules. For insoluble compounds, the energy released during hydration is less than the energy absorbed.

14.23 As NaCl dissolves in water, the attractive forces holding the ions together must be overcome. Each positive Na^+ ion is attracted to the negative O of water, and each negative Cl^- ion is attracted to a positive H of water.

14.25 a. soluble

b. insoluble (dolomite does react with acids to give CO_2 and soluble HCO_3^- ions)

c. all nitrates are soluble

d. insoluble

14.27 Dissolving and crystallizing occur at equal rates in a saturated solution.

14.29 a. The amount of KNO_3 that will dissolve in 200. g of water at 70 °C is

$$200.\text{ g } H_2O \times \frac{135 \text{ g } KNO_3}{100. \text{ g } H_2O} = 270. \text{ g } KNO_3$$ can dissolve. None of the KNO_3 present (200. g) will settle out (crystallize).

b. The amount of KNO_3 that will dissolve in 200. g of water at 10 °C is

$$200.\text{ g } H_2O \times \frac{20 \text{ g } KNO_3}{100. \text{ g } H_2O} = 40. \text{ g } KNO_3$$ can dissolve. The amount that will crystallize out is 200. g– 40. g = 160. g KNO_3.

14.31 $350.\text{ g water} \times \frac{36 \text{ g NaCl}}{100.\text{ g water}} = 126 \text{ g NaCl}$

14.33 Most soluble solids become more soluble with increasing temperatures because the motion of particles increases at higher temperatures, allowing more particles to be broken loose from the crystal. Once free, the particles in solution move too rapidly to be recaptured by the solid. Gases become less soluble with increasing temperatures because the rapidly moving gaseous molecules escape from the liquid surface in an open container.

14.35 The warm glass of soft drink will go "flat" more quickly because a gas is less soluble at higher temperatures where molecules move about more rapidly.

14.37 The solubility of NaCl (Figure 14.4) is about 38 g / 100 g water, regardless of temperature. The quantity that will dissolve in 175 g of water is $175 \text{ g } H_2O \times \frac{38 \text{ g NaCl}}{100. \text{ g } H_2O} = 66 \text{ g NaCl}$, so yes, 50. g will dissolve.

14.39 Both a and b are supersaturated with sugar, c is not supersaturated.

14.41 The calculation plan: mL $\rightarrow$ L $\rightarrow$ mol sucrose $\rightarrow$ g sucrose

$$0.500 \text{ L} \times \frac{0.10 \text{ mol}}{\text{L}} \times \frac{342 \text{ g sucrose}}{\text{mol}} = 17 \text{ g sucrose}$$

Dissolve the 17 g of sucrose in enough water to make a total volume of 500. mL of solution.

14.43 The calculation plan: mol sucrose → L solution → mL solution

$$0.040 \text{ mol} \times \frac{1 \text{ L}}{0.100 \text{ mol}} \times \frac{1000 \text{ mL}}{1 \text{ L}} = 400. \text{ mL}$$

14.45 The calculation plan: g sucrose → mol sucrose → L solution → mL solution

$$8.00 \text{ g sucrose} \times \frac{1 \text{ mol}}{342 \text{ g sucrose}} \times \frac{1 \text{ L}}{0.100 \text{ mol}} \times \frac{1000 \text{ mL}}{1 \text{ L}} = 234 \text{ mL}$$

14.47 $$\frac{\text{Volume of solution}}{\text{Total volume of solution}} \times 100\ \% = \text{percent by volume}$$

$$\frac{200. \text{ mL}}{500. \text{ mL}} \times 100\ \% = 40.0\,\% \text{ ethanol by volume}$$

Ethanol is the solute; water is the solvent.

14.49 60.0% (v/v) alcohol = 60.0 mL of alcohol/100 mL solution

$$500. \text{ mL solution} \times \frac{60.0 \text{ mL alcohol}}{100. \text{ mL solution}} = 300.0 \text{ mL alcohol}$$

Add enough water to 300. mL of alcohol to give 500. mL of solution.

14.51 10.0% (v/v) ethanol = 10.0 mL of ethanol/100 mL solution

$$750. \text{ mL solution} \times \frac{10.0 \text{ mL ethanol}}{100. \text{ mL solution}} = 75.0 \text{ mL ethanol}$$

14.53 a. The mass of solute (sucrose) needed is 0.25% x 800. g = 0.0025 x 800. g = 2.0 g sucrose, the solute. The mass of water needed is 800. g total – 2.0 g solute = 798 g water.

b. Use the same quantities of solute and water as shown in part a.

c. The mass of NaCl needed is 0.92% x 5.0 kg = 0.0092 x 5000 g = 46 g NaCl, the solute. The mass of water needed is 5000 g – 46 g solute = 4954 g water.

14.55 70.0% HNO_3 by mass is 70.0 g HNO_3/100. g acid

$$500. \text{ g acid} \times \frac{70.0 \text{ g } HNO_3}{100. \text{ g acid}} = 350. \text{ g } HNO_3$$

14.57 We will need the equality: 1 ppt = 1000 ppm

$$0.015 \text{ ppm} \times \frac{1 \text{ ppt}}{1000 \text{ ppm}} = 1.5 \times 10^{-5} \text{ ppt}$$

14.59 We will need the equality: 1 ppm = 1000 ppb

$$5.0 \text{ ppb} \times \frac{1 \text{ ppm}}{1000 \text{ ppb}} = 5.0 \times 10^{-3} \text{ ppm}$$

14.61 We will need the equality: 1 ppm = 1 mg/L

$$0.05 \text{ ppm} \times \frac{1 \text{ mg/L}}{1 \text{ ppm}} = 0.05 \text{ mg/L}$$

We will need the equality: 1 ppm = 1000 ppb

$$0.05 \text{ ppm} \times \frac{1000 \text{ ppb}}{1 \text{ ppm}} = 50 \text{ ppb}$$

14.63 $$\frac{0.10 \text{ mol}}{\text{L}} \times \frac{46 \text{ g}}{\text{mol}} = 4.6 \text{ g/L}$$

$$\frac{4.6 \text{ g ethanol}}{\text{L}} \times \frac{1 \text{ mL ethanol}}{0.785 \text{ g}} \times \frac{1 \text{ L ethanol}}{1000 \text{ mL ethanol}} \times 100\% = 0.58\% \text{ (v/v)}$$

14.65 $V_1C_1 = V_2C_2$ $\quad V_1 \times 70.0\ \% = 350.\ \text{mL} \times 40.0\ \%$

$$V_1 = \frac{350.\ \text{mL} \times 40.0\%}{70.0\%} = 200.\ \text{mL}$$ of antiseptic is diluted with 150 mL water

14.67 $V_1C_1 = V_2C_2$ $\quad V_1 \times 15.0\ \text{M} = 100.\ \text{mL} \times 6.00\ \text{M}$

$$V_1 = \frac{100.\ \text{mL} \times 6.00\ \text{M}}{15.0\ \text{M}} = 40.0\ \text{mL}$$ of solution is diluted with water to give a total volume of 100. mL

14.69 $V_1C_1 = V_2C_2$ $\quad 120\ \text{mL} \times M_1 = 50.0\ \text{mL} \times 3.00\ \text{M}$

$$M_1 = \frac{50.0\ \text{mL} \times 3.00\ \text{M}}{120.\ \text{mL}} = 1.25\ \text{M}$$

14.71 The 5 kg ethanol is 5000 g x 1 mol/46 g = 109 mol.

The 5 kg of methanol is 5000 g x 1 mol/32 g = 156 mol.

Each molecule of ethanol or of methanol gives only one particle in solution, but we have a greater number of dissolved moles (and of molecules of methanol), so the methanol sample is the more effective antifreeze.

14.73 a. 3 b. 2 c. 4 d. 1

14.75 Particles in solution have diameters of about 0.1 to 1.0 nm. Particles in suspensions have diameters of 100 nm or more. Colloidal particles have intermediate diameters (about 1.0 to 100 nm).

14.77 a. solid emulsion; liquid; solid b. liquid emulsion; liquid; liquid c. solid foam; gas; solid d. aerosol; liquid; gas

14.79 Crenation: The water in the cell will flow to the concentrated salt solution and the cells would shrink and die.

14.81 An isotonic solution is one that exhibits the same osmotic pressure as that of the fluid inside of a living cell. A hypotonic solution contains a lower concentration of dissolved particles than the cell. A hypertonic solution contains a higher concentration of dissolved particles than the cell.

14.83 The nasal spray is hypotonic. The concentration is less than the 0.92% concentration required for an isotonic solution.

14.85 The carbonated beverage goes "flat" at a higher temperature more quickly because gases are less soluble in liquids at higher temperatures.

14.87 a. The solute is NaCl, the solvent is water, and the solution is a liquid.

b. The solute is Cu, the solvent is Ag, and the solution is a solid.

c. The solute is I_2, the solvent is alcohol, and the solution is a liquid.

14.89 $$800.\text{ g acid} \times \frac{95\text{ g } H_2SO_4}{100.\text{ g acid}} = 760.\text{ g } H_2SO_4$$

$$800.\text{ g } H_2SO_4 \times \frac{100.\text{ g acid}}{95\text{ g } H_2SO_4} \times \frac{1\text{ mL acid}}{1.84\text{ g acid}} = 458\text{ mL acid}$$

14.91 $$\text{Percent by mass} = \frac{\text{mass of solute}}{\text{total mass of solution}} \times 100\ \% = \frac{100.\text{ g solute}}{500.\text{ g total}} \times 100\% = 20.0\%\ (w/w)$$

14.93 For aqueous solutions, a concentration of 1 mg/L is equivalent to 1 ppm so 0.1 mg/L is 0.1 ppm.

$$0.1\text{ ppm} \times \frac{1000\text{ ppb}}{1\text{ ppm}} = 100\text{ ppb}$$

14.95 During osmosis, only the solvent molecules pass through a semipermeable membrane. During dialysis, the semipermeable membrane allows passage of ions and small solute molecules along with the solvent, but retains large molecules and colloidal particles.

14.97 The crispness of vegetables can be lost when they are boiled in water to which salt has been added because water flows across the cell membrane out of the vegetable cells and into the more concentrated salt solution.

14.99 A phenomenon, called the Tyndall effect, occurs when the path of a beam of light is visible when passed through a liquid or gas (such as air). The effect is caused by light being scattered by colloidal particles. This phenomenon can often be observed by looking sideways at the beam of light from a movie projector or the beam of a strong spotlight or searchlight.

14.101 If a tomato is placed in a brine solution the water will move from the tomato cells into the brine solution. The water can then be removed from the brine solution in outdoor solar evaporation units.

14.103 A colloid has particle sizes from 1 to 100 nm and a suspension has particles with diameters greater than 100 nm.

14.105 $(1000.\text{ mL})(M_1) = (250.\text{ mL})(6.00\text{ M})$ $M_1 = 1.05\text{ M}$

14.107 $$0.425\text{ g NaCl} \times \frac{100.\text{ g solution}}{5.00\text{ g NaCl}} = 8.50\text{ g NaCl soluton}$$

CHAPTER 15
Reaction Rates and Chemical Equilibrium

15.1 The three factors are collision frequency, orientation, and activation energy.

15.3 Particles must get together (collide) before they can react. When the rate of collision increases, the reaction rate also increases.

15.5 The specific atom of a molecule that is to be transferred to a second atom or molecule must be at the site of the collision. The rate of reaction increases when colliding particles have the required collision geometry.

15.7 *Activation energy* is the minimum kinetic energy that colliding particles must have—or must overcome—in order for a reaction to occur.

15.9 The activation energy for the oxidation of glucose is much less than the activation energy for the photosynthesis of glucose.

15.11 At higher temperatures reaction rates increase because particles that move more slowly will collide less frequently. Also, the slower moving particles may not have the minimum activation energy for effective collisions.

15.13 Meat keeps longer in the freezer where temperatures are colder than in a refrigerator because the biochemical reactions that take place when food spoils are slowed at the lower temperature.

15.15 At low temperatures, chemical reactions occur more slowly, so at the low temperatures of hibernating animals, metabolism—the chemical reactions involved in the utilization of food—must occur more slowly. Because metabolism is slowed, smaller amounts of food are utilized.

15.17 When the surface area is increased the reaction rate is also increased. This is because the greater surface area allow reacting particles to collide more frequently.

15.19 A *catalyst* is a substance that speeds up the rate of chemical reaction without itself being consumed in the reaction. MnO_2 is one of the catalysts that can be used for the decomposition of hydrogen peroxide.

15.21 A catalyst lowers the activation energy for a reaction—by providing an alternate reaction mechanism—and therefore speeds up the reaction rate.

15.23 Sugar does not react with oxygen in the air at room temperature, but it is metabolized in living cells by a series of small steps, each of which requires a specific enzyme as a catalyst.

15.25 MnO_2 is a catalyst that speeds up the decomposition of H_2O_2.

15.27 MnO_2 lowers the activation energy for the decomposition of H_2O_2.

15.29 a. Chlorine atoms react with ozone to give ClO molecules and O_2 gas.

b. ClO molecules react with oxygen atoms to give Cl atoms and O_2 gas.

c. The Cl atom that reacts with ozone in step 1 is released in step 2 and is free to repeat the cycle again and again by reacting with—and decomposing—more ozone molecules.

15.31 In this model, dynamic equilibrium is established when the number of people going up the escalator equals the number of people going down the escalator per minute or another unit of time. The number of people on one floor could be quite large compared with the number of people on the other floor. Similarly, equilibrium does not mean that moles of reactants equals moles of products, but that the forward and reverse rates are equal and concentrations (numbers per floor) remain constant.

15.33 Early in the morning the parking lot may be almost empty. Soon, vehicles begin to arrive. Dynamic equilibrium is established when the number of vehicles entering the parking lot per unit time equals the number of vehicles that leave; thus, the concentration of vehicles in the lot becomes constant.

15.35 a. Physical equilibrium. Melting (and freezing) point.

b. Physical equilibrium. Equilibrium occurs when a saturated solution is present with excess sodium chloride.

c. Chemical equilibrium. Equilibrium is established when the rate of the forward reaction equals the rate of the reverse reaction.

15.37 Molecules of reactants are being converted to molecules of products and vice versa, but both processes occur at the same rate.

15.39 Before equilibrium is established, the concentration of the reactant (A) decreases while the concentration of product (B) increases.

15.41 When equilibrium is established—in this analogy—the number of uncombined nuts and bolts is stabilized (constant) and the number of combined nuts and bolts is also stabilized (constant).

15.43 At equilibrium in the nuts and bolts analogy, the rate at which nuts and bolts are assembled—the forward reaction—is equal to the rate at which nuts and bolts are taken apart—the reverse reaction.

15.45 Before equilibrium is established, the concentrations of reactants, CO and O_2, decrease as the concentration of CO_2 increases. When equilibrium is established, the concentrations of CO, O_2, and CO_2 remain constant.

15.47 a. Equilibrium shifts toward the left to favor (increase) the production of more reactants.

b. Equilibrium shifts toward the right to favor (increase) the production of more products.

c. Equilibrium shifts toward the right to favor (increase) the production of more products.

15.49 a. Increasing total pressure will have no effect on the position of the equilibrium (because the number of moles of gas does not change).

b. Increasing total pressure will shift equilibrium to the left to favor an increase in concentrations of reactants.

c. Increasing total pressure will shift equilibrium to the right to favor an increase in concentrations of products.

15.51 a. The equilibrium shifts to the right to favor an increase in concentrations of products.

b. The equilibrium shifts to the right to favor an increase in concentrations of products.

c. The equilibrium shifts to the right to favor an increase in concentrations of products.

d. The use of a catalyst does not affect the position of the equilibrium.

15.53 a. The equilibrium shifts to the right to favor an increase in concentrations of products.

b. The equilibrium shifts to the left to favor an increase in concentrations of reactants.

15.55 A catalyst increases the rate of a reaction by lowering the activation energy. With a lower activation energy, a greater fraction of reactants have sufficient energy to react, so the rate of the reaction increases.

15.57 a. The equilibrium shifts to the right, favoring methanol production.

b. The equilibrium shifts to the right, favoring methanol production.

c. The equilibrium shifts to the left to favor an increase in concentrations of reactants.

d. The use of a catalyst does not affect the position of the equilibrium.

15.59 The temperature of the exhaust gases is not great enough for high percentages of the gases to react. A catalyst is needed so the reactions can proceed more efficiently at lower activation energies.

15.61 a. $K_{eq} = \frac{[HCl]^2}{[H_2][Cl_2]}$ b. $K_{eq} = \frac{[CO]^2[O_2]}{[CO_2]^2}$ c. $K_{eq} = \frac{[O_3]^2}{[O_2]^3}$

15.63 a. $K_{eq} = \frac{[SO_3]^2}{[SO_2]^2[O_2]}$ b. $K_{eq} = \frac{[N_2O_4]}{[NO_2]^2}$ c. $K_{eq} = [CO_2]$

15.65 The equilibrium lies to the right; the concentrations of products are greater than concentrations of reactants.

15.67 a. To the right, favoring products.

b. To the left, favoring reactants.

c. To the right, favoring products

d. To the left, favoring reactants, but the K_{eq} is too close to 1 to tell for sure.

15.69 Equilibrium lies far to the right. See Section 15.6. When K_{eq} is much larger than one (positive exponent) equilibrium lies to the right because large concentrations of products (numerator) divided by small concentrations of reactants (denominator) gives a constant that is greater than one.

15.71 $K_{eq} = \frac{[NO_2]^2}{[N_2O_4]} = 0.212$

$$\frac{(0.20)^2}{[N_2O_4]} = 0.212$$

$$[N_2O_4] = \frac{(0.20)^2}{0.212} = 0.19\ M$$

15.73 The K_{eq} is not affected by changes in concentration.

15.75 a. $K_{eq} = \frac{[CO_2][H_2]}{[CO][H_2O]}$ b. $3.72 = \frac{[1.18][H_2]}{[0.621][0.668]}$ $H_2 = \frac{3.72[0.621][0.668]}{[1.18]} = 1.31\ M$

15.77 a. The equilibrium shifts to the right, favoring the products.

b. The equilibrium shifts to the right, favoring the products.

c. The use of a catalyst does not affect the position of the equilibrium.

15.79 a. To the left, favoring reactants.

b. To the right, favoring products.

c. A catalyst does not effect the equilibrium.

d. To the right, favoring products.

e. $K_{eq} = \frac{[SO_2]^2[O_2]}{[SO_3]^2}$

$0.234 = \frac{[SO_2]^2[0.205M]}{[0.623M]^2}$ $[SO_2]^2 = \frac{0.234 \text{ x } (0.632M)^2}{0.205M}$ $[SO_2]^2 = 0.443\ M^2$ $[SO_2] = 0.666$

15.81 In the high temperature and pressure present in an automobile engine, some nitrogen gas from the air combines with oxygen from the air to give oxides of nitrogen. Certain catalysts are used in a catalytic converter to decompose these oxides of nitrogen to give N_2 and O_2.

15.83 a. $K_{eq} = \frac{[NO]^2}{[N_2]\,[O_2]}$

b. Higher engine temperatures tend to increase NO(g) pollution because adding heat shifts the equilibrium to the right for this reaction.

15.85 The equilibrium concentration of H_2S is very large compared to concentrations of products. See section 15.6. When K_{eq} is much smaller than one (negative exponent) equilibrium lies to the left because small concentrations of products (numerator) divided by large concentrations of reactants (denominator) gives a constant that is less than one.

15.87 The NO(g) concentration will be greater at 2500 K. Both K_{eq} values are less than one, but the K_{eq} value is larger at 2500 K than at 2000 K, and the larger the constant, the greater the concentrations of products.

15.89 $NO_2(g)$ increases because with the added heat the equilibrium is shifted to the left.

15.91 a. $[O_3]$ increases. The reaction is endothermic, so a higher temperature shifts the equilibrium to the right, favoring an increase in $[O_3]$.

b. Equilibrium shifts to the right to replace some of the ozone that was removed. Equilibrium shifts to at least partially restore what was removed.

c. Equilibrium shifts to the right to favor products, where there are fewer moles of gas.

15.93 a. Equilibrium shifts to the right to favor products, having fewer moles of gas.

b. Equilibrium shifts to the right. Adding a reactant increases product concentration.

15.95 a. No effect. The number of moles of gas for reactants and products does not change.

b. No effect. A catalyst increases the rate of both forward and reverse reactions but does not change equilibrium concentrations.

c. [NO] increases. With a negative ΔH the reaction is exothermic. A higher temperatureshifts the equilibrium to the left, favoring an increase in [NO].

15.97 $[SO_3]^2 = K_{eq}[SO_2]^2[O_2] = 2.80 \times 10^2[0.00880]^2[0.0119]$, $[SO_3] = 0.0161$

15.99 $[NO_2] = K_{eq}[N_2][O_2] = 2.10 \times 10^{-3}[3.48 \times 10^{-3}][9.38 \times 10^{-3}] = 6.85 \times 10^{-7}$, $[NO] = 8.28 \times 10^{-4}$ M

CHAPTER 16
Acids and Bases

16.1 (1) Acids taste sour. (2) Acids cause litmus to change from blue to red. (3) Acids react with active metals to produce hydrogen gas. (4) Acids react with bases to form water and salts.

16.3 Arrhenius attributed properties of acids to the presence of hydrogen ions.

16.5 A hydronium ion, H_3O^+, is a hydrogen ion, H^+, bonded to a water molecule. The terms are used interchangeably because all hydrogen ions are actually hydrated (they are joined to water molecules) to form hydronium ions.

16.7 a. acid b. base c. salt d. acid e. none of these f. salt

16.9 a. CH_3COOH (monoprotic) b. H_3BO_3 (triprotic)

c. HCl (monoprotic) d. H_2SO_4 (diprotic)

16.11 Three strong bases: NaOH (sodium hydroxide), KOH (potassium hydroxide), and $Ca(OH)_2$ (calcium hydroxide). Aqueous ammonia, NH_3(aq) is a weak base.

16.13 Strong acids are those that ionize completely in water; weak acids do not.

16.15 A weak acid is one that ionizes only to a small extent in water. After water is mixed with a concentrated acid, it is called a dilute acid.

16.17 Aqueous ammonia is NH_3 dissolved in water. A relatively small amount of the NH_3 reacts with water to produce OH^- ions and NH_4^+ ions in solution.

16.19 a. weak acid b. strong acid c. weak acid

d. weak acid e. strong base f. strong base

16.21 Yes, many oven and drain cleaners contain strong bases. Muratic acid, HCl is a strong acid and can damage your eyes. Vinegar is weak acid.

16.23 Although the percent of acid is the same for both acids, the sulfuric acid is a strong acid that is totally ionized, while the vinegar contains a weak acid (acetic acid) that is only slightly ionized.

16.25 $H_2SO_4(aq) + 2\ KOH \rightarrow K_2SO_4(aq) + 2\ H_2O$

The net ionic equation: $H^+ + OH^- \rightarrow H_2O$

16.27 a. $H_2SO_4\ (aq) + ZnO \rightarrow ZnSO_4\ (aq) + H_2O\ (g)$

b. $H_2SO_4\ (aq) + Zn \rightarrow ZnSO_4\ (aq) + H_2\ (g)$

c. $H_2SO_4\ (aq) + CaCO_3 \rightarrow CaSO_4\ (s) + CO_2\ (g) + H_2O$

16.29 Amphoteric metals are those that react directly with strong bases and water to produce hydrogen gas. Examples include aluminum and zinc.

16.31 a. Already balanced. $Al(OH)_3$ will react with a strong base.

b. $Al(OH)_3 + 3\ HCl(aq) \rightarrow AlCl_3(aq) + 3\ H_2O$

$Al(OH)_3$ will react with a strong acid and a strong base. $Al(OH)_3$ is an amphoteric hydroxide.

16.33 a. $HCl(aq) + NaHCO_3 \rightarrow NaCl(aq) + CO_2(g) + H_2O$

the net ionic equation : $H^+ + HCO_3^- \rightarrow CO_2 + H_2O$

16.35 An acid, according to Brønsted-Lowry, is a proton donor. According to the Lewis definition, an acid is an electron pair acceptor.

16.37 Water is an acid: $NH_3(aq) + H_2O \rightarrow NH_4^+(aq) + OH^-(aq)$

16.39 $HCl(g) + NH_3(g) \rightarrow NH_4Cl(s)$ This is an acid-base reaction by Brønsted-Lowry and Lewis definitions, but is not according to Arrhenius.

16.41 a. C_6H_5SH is a Brønsted acid (it donates a proton). $C_6H_5S^-$ is its conjugate base.

b. CH_3NH_2 is a Brønsted base (it accepts a proton). $CH_3NH_3^+$ is its conjugate acid.

c. NH_3 is a Brønsted base that forms NH_4^+, its conjugate acid.

16.43 a. HBr and HF are the acids. HBr is the stronger acid of the two; it gives up a proton more readily so there are more Br^- ions in the solution than F^- ions in solution.

b. HCN and HF are the acids. HF is the stronger of the two; it gives up a proton more readily so there are more F^- ions in solution than CN^- ions in solution.

16.45 a. F^- and Br^- are bases. F^- is the stronger base of the two; it accepts a proton more readily so there is more HF than HBr in solution.

b. F^- and CN^- are bases. CN^- is the stronger base of the two; it accepts a proton more readily so there is more HCN than HF in solution.

16.47 Aniline, $C_6H_5NH_2$, acts as a base, accepting a proton to form $C_6H_5NH_3^+$.

16.49 a. $K_w = [H^+][OH^-] = 1.0 \times 10^{-14}$, $[1.0 \times 10^{-3}][OH^-] = 1.0 \times 10^{-14}$
$[OH^-] = 1.0 \times 10^{-14}/1.0 \times 10^{-3} = 1.0 \times 10^{-11}$
To find pH, substitute the $[H^+]$ in the mathematical expression for pH.
When the coefficient of 10^{-n} is exactly 1, we do not need a calculator. The pH is equal to the number, n, in the exponent. $pH = -\log[H^+] = -\log(1.0 \times 10^{-3}) = 3.0$
Answer: The pH is 3.0, which is less than 7, so the sample is acidic.

b. $K_w = [H^+][OH^-] = 1.0 \times 10^{-14}$, $[1.0 \times 10^{-9}][OH^-] = 1.0 \times 10^{-14}$
$[OH^-] = 1.0 \times 10^{-14}/1.0 \times 10^{-9} = 1.0 \times 10^{-5}$
To find pH, substitute the $[H^+]$ in the mathematical expression for pH.
When the coefficient of 10^{-n} is exactly 1, we do not need a calculator. The pH is equal to the number, n, in the exponent. $pH = -\log[H^+] = -\log(1.0 \times 10^{-9}) = 9.0$
Answer: The pH is 9.0, which is greater than 7, so the sample is basic.

c. $K_w = [H^+][OH^-] = 1.0 \times 10^{-14}$, $[H^+][1.0 \times 10^{-5}] = 1.0 \times 10^{-14}$
$[H^+] = 1.0 \times 10^{-14}/1.0 \times 10^{-5} = 1.0 \times 10^{-9}$
To find pH, substitute the $[H^+]$ in the mathematical expression for pH.
When the coefficient of 10^{-n} is exactly 1, we do not need a calculator. The pH is equal to the number, n, in the exponent. $pH = -\log[H^+] = -\log(1.0 \times 10^{-9}) = 9.0$
Answer: The pH is 9.0, which is greater than 7, so the sample is basic.

d. $K_w = [H^+][OH^-] = 1.0 \times 10^{-14}$, $[H^+][1.0 \times 10^{-10}] = 1.0 \times 10^{-14}$
$[H^+] = 1.0 \times 10^{-14}/1.0 \times 10^{-10} = 1.0 \times 10^{-4}$
To find pH, substitute the $[H^+]$ in the mathematical expression for pH.
When the coefficient of 10^{-n} is exactly 1, we do not need a calculator. The pH is equal to the number, n, in the exponent. $pH = -\log[H^+] = -\log(1.0 \times 10^{-4}) = 4.0$
Answer: The pH is 4.0, which is less than 7, so the sample is acidic.

16.51 a. $pH = 14 - pOH = 14 - 10.00 = 4.00$

b. $pH = 14 - pOH = 14 - 8.6 = 5.4$

c. $pOH = 14 - pH = 14 - 5.00 = 9.00$

d. $pOH = 14 - pH = 14 - 10.3 = 3.7$

16.53 a. $[H^+] = 1.00 \times 10^{-6}$ M $pH = -\log[H^+] = -\log(1 \times 10^{-6}) = 6.000$

b. $[OH^-] = 1.00 \times 10^{-5}$ $[H^+][OH^-] = 1.00 \times 10^{-14}$ $[H^+] = 1.0 \times 10^{-14}/1.0 \times 10^{-5} = 1.0 \times 10^{-9}$ M
$pH = -\log[H^+] = -\log(1.0 \times 10^{-9}) = 9.000$

c. $[H^+] = 1.00 \times 10^{-2}$ M $pH = -\log[H^+] = -\log(1 \times 10^{-2}) = 2.000$

d. $[OH^-] = 1.00 \times 10^{-2}$ $[H^+][OH^-] = 1.00 \times 10^{-14}$ $[H^+] = 1.0 \times 10^{-14}/1.0 \times 10^{-2} = 1.0 \times 10^{-12}$ M
$pH = -\log[H^+] = -\log(1.0 \times 10^{-12}) = 12.000$

16.55 a. To find pH, substitute the $[H^+]$ in the mathematical expression for pH.
To find the logarithm of 3.4×10^{-5}, use a calculator with a LOG key and follow the instructions given in Example 16.12. $pH = -\log(3.4 \times 10^{-5}) = 4.47$

b. $pH = -\log(7.2 \times 10^{-12}) = 11.14$

c. $[OH^-] = 8.5 \times 10^{-7}$ $[H^+][OH^-] = 1.00 \times 10^{-14}$ $[H^+] = 1.0 \times 10^{-14}/18.5 \times 10^{-7} = 1.2 \times 10^{-8}$ M
$pH = -\log[H^+] = -\log(1.2 \times 10^{-8}) = 7.93$

d. $[OH^-] = 6.3 \times 10^{-6}$ $[H^+][OH^-] = 1.0 \times 10^{-14}$ $[H^+] = 1.0 \times 10^{-14}/6.3 \times 10^{-6} = 1.59 \times 10^{-9}$ M
$pH = -\log[H^+] = -\log(1.3 \times 10^{-5}) = 8.80$

16.57 a. $pH = -\log[H^+]$ $8.35 = -\log[H^+]$ $[H^+] = \text{antilog}(-8.35) = 4.5 \times 10^{-9}$ M

b. $pH = -\log[H^+]$ $2.73 = -\log[H^+]$ $[H^+] = \text{antilog}(-2.73) = 1.9 \times 10^{-3}$ M

c. $pH = 14 - pOH = 14 - 9.10 = 4.90$
$pH = -\log[H^+]$ $4.90 = -\log[H^+]$ $[H^+] = \text{antilog}(-4.90) = 1.3 \times 10^{-5}$ M

d. $pH = 14 - pOH = 14 - 6.08 = 7.92$
$pH = -\log[H^+]$ $7.92 = -\log[H^+]$ $[H^+] = \text{antilog}(-7.92) = 1.2 \times 10^{-8}$ M

16.59 a. Acidic; this is a salt of a weak base, $Mg(OH)_2$, and a strong acid, HBr.

b. Close to neutral; this is a salt of a weak base, NH_3 and a weak acid, H_3PO_4.

c. Basic; this is a salt of a strong base, NaOH, and a weak acid, H_2CO_3.

d. Basic; this is a salt of a strong base, NaOH, and a weak acid, CH_3COOH.

e. Basic; this is a salt of a strong base, NaOH and a weak acid, HCN.

16.61 Acidic; this is a salt of a weak base, NH_3, and a strong acid, HNO_3.

16.63 A buffer is prepared by using a weak acid and a salt of that acid, or a weak base and a salt of that base.

16.65 Aqueous ammonia; it is the weak base that gives aqueous ammonium ions.

16.67 When H^+ ions from an acid enter the blood, the H^+ ions react with HCO_3^- ions in the blood to form carbonic acid, H_2CO_3, a weak acid.

$H^+(aq) + HCO_3^-(aq) \rightarrow H_2CO_3(aq)$

16.69 In an acid-base titration, the concentration of an acid can be determined by adding just enough base, dropwise, to neutralize the acid. At the end point, the concentration of the acid can be calculated when the concentration of the base and volumes of both the acid and the base are known.

16.71 Acidic; the salt formed is that of a strong base and a weak acid.

16.73 STEP 1 Write a balanced equation for the neutralization reaction.

$HCl + KOH \rightarrow KCl + H_2O$

STEP 2 List the volume and concentration for the acid and base.

	Acid	Base
Volume	20.00 mL	10.00 mL
Molarity	? M	0.5000 M

STEP 3 The conversion needed:

Volume of base → Moles of base → Moles of acid → M of acid

STEP 4 Follow through with the conversions:

$$0.01000 \text{ L base} \times \frac{0.5000 \text{ mol base}}{\text{L base}} \times \frac{1 \text{ mol acid}}{1 \text{ mol base}} = 0.005000 \text{ mol acid}$$

STEP 5 Divide mole of acid by liters of acid to obtain the molarity, M.

$$\frac{0.005000 \text{ mol acid}}{0.02000 \text{ L acid}} = 0.2500 \text{ M HCl}$$

16.75 STEP 1 Write a balanced equation for the neutralization reaction.

H_2SO_4 + 2 NaOH → $2H_2O$ + Na_2SO_4

STEP 2 List the volume and concentration for the acid and base

	Acid	Base
Volume	12.53 mL	31.22 mL
Molarity	0.1000 M	? M

STEP 3 The conversion needed:

Volume of acid → Moles of acid → Moles of base → M of base

STEP 4 Follow through with the conversions:

$$0.01253 \text{ L acid} \times \frac{0.1000 \text{ mol acid}}{\text{L acid}} \times \frac{2 \text{ mol base}}{1 \text{ mol acid}} = 0.002506 \text{ mol base}$$

STEP 5 Divide mole of acid by liters of acid to obtain the molarity, M.

$$\frac{0.002506 \text{ mol base}}{0.03122 \text{ L base}} = 0.08027 \text{ M NaOH}$$

16.77 There is a ratio of 1 mole of acid to two moles of base. Find the number of moles of base required to react with the given amount of acid and compare.

$$0.098 \text{ L base} \times \frac{6.0 \text{ mol base}}{\text{L base}} \times \frac{1 \text{ mol acid}}{2 \text{ mol base}} = 0.294 \text{ mole acid required}$$

$$0.150 \text{ L acid} \times \frac{2.0 \text{ mol acid}}{\text{L acid}} = 0.300 \text{ mole acid available}$$

There are 0.300 – 0.294 = 0.006 moles of excess acid. Divide the number of moles of excess acid by the total volume to determine the molarity of the final solution. 0.006 mole/ 0.248 L = 0.024 M acid.
pH = -log(0.024) = 1.6

16.79 a. diprotic (two acidic H's) b. monoprotic (one acidic H)

c. monoprotic (one acidic H) d. triprotic (three acidic H's)

16.81 $CaCO_3$ + H_2SO_4 → $CaSO_4$ + CO_2 + H_2O

16.83 $NaHCO_3$ + CH_3COOH → $NaCH_3COO$ + CO_2 + H_2O

16.85 Calcium carbonate, magnesium hydroxide, and aluminum hydroxide.

16.87

HF	+ H_2O	⇔	H_3O^+	+	F^-	
weaker acid	weaker base		stronger acid		stronger base	HF/F^- and H_3O^+/H_2O are conjugate acid/base pairs

16.89 $Ca(OH)_2$ + H_2SO_4 → $CaSO_4$ + 2 H_2O
Thus one mole of $Ca(OH)_2$ can neutralize 1 mole of H_2SO_4.

$$10 \text{ tons Ca(OH)}_2 \times \frac{1 \text{ mol Ca(OH)}_2}{74.1 \text{ g Ca(OH)}_2} \times \frac{1 \text{ mol } H_2SO_4}{1 \text{ mol Ca(OH)}_2} \times \frac{98.0 \text{ g } H_2SO_4}{1 \text{ mol } H_2SO_4} = 13.2 \text{ tons } H_2SO_4$$

16.91 This answer depends on the antacids you select.

16.93 a. Brønsted-Lowry and Lewis definitions apply. The N_2H_4 is a Brønsted base (it accepts a proton) and a Lewis base (it donates an electron pair).

b. Only the Lewis definition applies here. The SO_3 is the Lewis acid.

16.95 a. To find the $[OH^-]$ use the K_w expression.
$[OH^-]=1.0 \times 10^{-14}/1.0 \times 10^{-4} = 1.0 \times 10^{-10}$
$pH = -\log[H^+] = -\log(1 \times 10^{-4}) = 4.00$
The pH is 4.00, which is less than 7, so the sample is acidic.

b. To find the $[OH^-]$ use the K_w expression.
$[OH^-]=1.0 \times 10^{-14}/8.78 \times 10^{-8} = 1.14 \times 10^{-7}$
$pH = -\log[H^+] = -\log(8.78 \times 10^{-8}) = 7.06$
The pH is 7.06, which is greater than 7, so the sample is slightly basic.

c. To find the $[H^+]$ use the K_w expression.
$[H^+]=1.0 \times 10^{-14}/1.0 \times 10^{-6} = 1.0 \times 10^{-8}$
$pH = -\log[H^+] = -\log(1 \times 10^{-8}) = 8.00$
The pH is 8.00, which is greater than 7, so the sample is basic.

d. To find the $[H^+]$ use the K_w expression.
$[H^+]=1.0 \times 10^{-14}/9.68 \times 10^{-9} = 1.03 \times 10^{-6}$
$pH = -\log[H^+] = -\log(1.03 \times 10^{-6}) = 5.99$
The pH is 5.99, which is less than 7, so the sample is acidic.

16.97 Ammonia is a weak base. It only partially hydrolyzes to produce OH^- ions. The solution should be referred to as a concentrated solution.

16.99 Muratic acid is a 38% solution of hydrochloric acid. Hydrochloric acid is a strong acid and completely ionizes. Muratic acid can cause severe burns.

16.101 a. W is more acidic due to greater $[H^+]$ b. 1000 times c. For W, pH = 3; for X, pH = 6

16.103 a. W is more basic due to greater $[OH^-]$ b. 100 times c. For W, pH = 13; for X, pH = 11

16.105 a. $[H^+] = 1.0 \times 10^{-3}$ M b. Original pH = 1; final pH = 3

16.107 a. $[OH^-] = 1.0 \times 10^{-4}$ M b. Original pH = 12; final pH = 10

16.109 a. NH_4^+ b. H_2O c. HNO_3

16.111 a. pH = 4.627; acidic b. pH = 8.057; basic c. pH = 11.810; basic d. pH = 4.7110; basic

CHAPTER 17
Oxidation and Reduction

17.1 a. In B_2O_3, the oxidation number of oxygen is –2 and there are three oxygen atoms. The sum must be 0. Let the oxidation number of boron be B.

$$2B + 3(-2) = 0$$
$$2B - 6 = 0$$
$$B = +3$$

The boron has an oxidation number of +3.

b. In $C_2O_4^{2-}$, the oxidation number of oxygen is –2 and there are 4 oxygen atoms. The sum must be –2. Let the oxidation number of carbon be represented by C in the equation.

$$2C + 4(-2) = -2$$
$$2C - 8 = -2$$
$$C = +3$$

The carbon has an oxidation number of +3.

c. Sulfur is an element, so its oxidation number is 0 by definition.

d. In ClO_4^-, the oxidation number of oxygen is –2 and there are 4 oxygen atoms. The sum must be –1. Let the oxidation number of chlorine be Cl.

$$Cl + 4(-2) = -1$$
$$Cl - 8 = -1$$
$$Cl = +7$$

The chlorine has an oxidation number of +7.

17.3 a. In Na_3P, the oxidation number of sodium is +1 and there are 3 sodium atoms. The sum must equal 0. Let the oxidation number of phosphorus be P.

$$3(+1) + P = 0$$
$$P = -3$$

The phosphorus has an oxidation number of –3.

b. In $P_2O_7^{4-}$, the oxidation number of oxygen is –2. The sum must equal –4. Let the oxidation number of phosphorus be P.

$$2P + 7(-2) = -4$$
$$2P = -4 + 14$$
$$P = +5$$

The phosphorus has an oxidation number of +5.

c. In H_3PO_4, the oxidation number of hydrogen is +1 and oxygen is –2. The sum must be 0. Let the oxidation number of phosphorus be P.

$$3(+1) + P + 3(-2) = 0$$
$$P - 3 = 0$$
$$P = +3$$

The phosphorus has an oxidation number of +3.

d. In $K_2H_2P_2O_7$, the oxidation number of potassium is +1, hydrogen is +1, and oxygen is –2. Let the oxidation number of phosphorus be P.

$$3(+1) + 2(+1)\ 2P + 7(-2) = 0$$
$$2P - 10 = 0$$
$$P = +5$$

The phosphorus has an oxidation number of +5.

17.5 a. In MnO_4^- the oxidation number of Mn is +7. The sum must be –1 and the oxidation number of O is -2.

$$Mn + 4(-2) = -1$$
$$Mn = +7$$

b. In MnO_2 the oxidation number of Mn is +4. The sum must be 0 and the oxidation number of O is -2.

$$Mn + 2(-2) = 0$$
$$Mn = +4$$

17.7 a. Oxidation occurs when an element or compound gains oxygen atoms.

b. Oxidation occurs when a compound loses hydrogen atoms.

c. Oxidation occurs when an atom or ion of an element loses electrons.

d. Oxidation occurs when there is an increase in oxidation number.

17.9 a. Water is reduced; it loses an oxygen atom.

b. Br_2 is reduced; bromine atoms gain electrons.

c. C_2H_4O is reduced; it gains hydrogen atoms.

d. C_2H_4O is oxidized; it gains oxygen atoms.

17.11 a. Not redox; there is no change in oxidation numbers.

b. Redox; there is a change in oxidation number for Cl and I.

c. Redox; there is a change in oxidation number for Cl and Pb.

d. Not redox; there is no change in oxidation numbers.

17.13 Reduced; the unsaturated oil gains hydrogen atoms.

17.15 a.
$$\underset{RA}{2\,Fe} + \underset{OA}{3\,Cl_2} \rightarrow 2\,FeCl_2$$ (Fe is oxidized to Fe^{3+})

b.
$$\underset{RA}{Mg} + \underset{OA}{Cu(NO_3)_2} \rightarrow Cu + Mg(NO_3)_2$$ (Mg is oxidized to Mg^{2+})

c.
$$\underset{OA}{2\,PbO} + \underset{RA}{C} \rightarrow 2\,Pb + CO_2$$ (C is oxidized to CO_2.)

d.
$$\underset{OA}{Cl_2} + \underset{RA}{2\,NaBr} \rightarrow Br_2 + 2\,NaCl$$ (Br^- ions are oxidized to Br_2)

17.17 a. S is oxidized. (S is oxidized from +4 in SO_2 to +6 in H_2SO_4.)

b. N is reduced. (N is reduced from +5 in HNO_3 to +4 in NO_2.)

c. HNO_3 is the oxidizing agent. (N atoms get reduced.)

d. SO_2 is the reducing agent. (S atoms get oxidized.)

17.19 a. Fe is oxidized. (Fe is oxidized from Fe^{2+} to Fe^{3+}.)

b. Mn is reduced. (Mn is reduced from +7 in MnO_4^- to +2 in Mn^{2+}.)

c. MnO_4^- is the oxidizing agent. (Mn atoms get reduced from +7 to +2.)

d. Fe^{2+} ions are reducing agents. (Fe^{2+} ions get oxidized to Fe^{3+}.)

17.21 a. Zr is oxidized. (Zr is oxidized from 0 to +4.)

b. H is reduced. (H is reduced from +1 to 0.)

c. H_2O is the oxidizing agent. (H atoms get reduced in H_2O.)

d. Zr is the reducing agent. (Zr atoms get oxidized.)

17.23 a. Indoxyl is oxidized. When two molecules of indoxyl combine to make one molecule of indigo dye, the total number of C and O atoms remains the same, but 4 H atoms are lost; they are oxidized to form water.

b. O_2 gas is the oxidizing agent. Oxygen combines with hydrogen—oxygen is reduced by hydrogen—to form water.

17.25 a. Cl_2 is the oxidizing agent. Its oxidation number changes from 0 to -1. Bromide ions are reducing agents. They are oxidized from -1 to 0.

b. $Cr_2O_7^{2-}$ ions are oxidizing agents. Chromium is reduced from +6 to +3. Iron(II) ions are reducing agents. They are oxidized from +2 to +3.

17.27 Reducing agents include H_2 gas, carbon (coke), and carbon monoxide gas.

17.29 Sulfur is oxidized. Its oxidation number changes from +4 in SO_3^{2-} ions to +6 in SO_4^{2-} ions.

17.31 Step 1. Write half-reactions

$$I^- \rightarrow I_2$$
$$SO_4^{2-} \rightarrow S$$

Step 2. Balance the I and S atoms.

$$2\,I^- \rightarrow I_2$$
$$SO_4^{2-} \rightarrow S$$

Step 3. Balance O atoms by using 4 H_2O.

$$2\,I^- \rightarrow I_2$$
$$SO_4^{2-} \rightarrow S + 4\,H_2O$$

Step 4. Balance H atoms by using 8 H^+.

$$2\,I^- \rightarrow I_2$$
$$8\,H^+ + SO_4^{2-} \rightarrow S + 4\,H_2O$$

Step 5. Balance electrical charges using 2 e^- and 6 e^-.

$$2\,I^- \rightarrow I_2 + 2\,e^-$$
$$6\,e^- + 8\,H^+ + SO_4^{2-} \rightarrow S + 4\,H_2O$$

Step 6. Electrons lost must equal electrons gained, so multiply the first equation by 3 and the second by 1. Add the half-reaction equations and cancel electrons (6 e^- on each side).

$$6\,I^- \rightarrow 3I_2 + 6\,e^-$$
$$6\,e^- + 8\,H^+ + SO_4^{2-} \rightarrow S + 4\,H_2O$$
$$6\,I^- + 8\,H^+ + SO_4^{2-} \rightarrow 3\,I_2 + S + 4\,H_2O$$

Step 7. Check to make sure all atoms and total charges on both sides are balanced and that all substances are in simplest integer ratios.

17.33 Step 1. Write half-reactions

$$Cl^- \rightarrow Cl_2$$
$$MnO_4^- \rightarrow Mn^{2+}$$

Step 2. Balance Cl atoms; Mn atoms are balanced.

$$2\,Cl^- \rightarrow Cl_2$$
$$MnO_4^- \rightarrow Mn^{2+}$$

Step 3. Balance O atoms by using 4 H_2O.

$$2\,Cl^- \rightarrow Cl_2$$
$$MnO_4^- \rightarrow Mn^{2+} + 4\,H_2O$$

Step 4. Balance H atoms by using 8 H^+.

$$2\,Cl^- \rightarrow Cl_2$$
$$8\,H^+ + MnO_4^- \rightarrow Mn^{2+} + 4\,H_2O$$

Step 5. Balance electrical charges using 2 e^- and 5 e^-.

$$2\,Cl^- \rightarrow Cl_2 + 2\,e^-$$
$$5\,e^- + 8\,H^+ + MnO_4^- \rightarrow Mn^{2+} + 4\,H_2O$$

Step 6. Electrons lost must equal electrons gained, so multiply the first equation by 5 and the second by 2. Add the half-reaction equations and cancel electrons (10 e^- on each side).

$$10\,Cl^- \rightarrow 5\,Cl_2 + 10\,e^-$$
$$10\,e^- + 16\,H^+ + 2\,MnO_4^- \rightarrow 2\,Mn^{2+} + 8\,H_2O$$
$$10\,Cl^- + 16\,H^+ + 2\,MnO_4^- \rightarrow 5\,Cl_2 + 2\,Mn^{2+} + 8\,H_2O$$

Step 7. Check to make sure all atoms and total charges on both sides are balanced and that all substances are in simplest integer ratios.

17.35 A voltaic cell (also called a galvanic cell) is an electrochemical cell that uses a spontaneous chemical reaction to generate an electrical current.

17.37 a. The equation represents reduction. (There is a loss of electrons.)

b. Pb is reduced. (It is reduced from +4 in PbO_2 to +2 in $PbSO_4$.)

c. This half-reaction occurs at the cathode. (Reduction always occurs at the cathode of every electrochemical cell.)

17.39 a. The equation represents reduction. (There is a loss of electrons.)

b. Ni is reduced. (It is reduced from +3 in Ni_2O_3 to +2 in $Ni(OH)_2$.)

c. This half-reaction occurs at the cathode. (Reduction always occurs at the cathode of every electrochemical cell.)

17.41 a. Oxidation. (There is a loss of electrons.) It occurs at the anode.

b. Reduction. (There is a gain of electrons.) It occurs at the cathode.

c. The net reaction is $2\,H_2(g) + O_2(g) \rightarrow 2\,H_2O$

17.43 a. The cathode (reduction) reaction is $Ca^{2+} + 2e^- \rightarrow Ca(s)$
(Reduction occurs at the cathode where there is a gain of electrons.)

b. The anode (oxidation) reaction is $2Br^- \rightarrow Br_2(g) + 2e^-$
(Oxidation occurs at the anode where there is a loss of electrons.)

17.45 A nickel metal strip (one electrode) is inserted into a solution of $Ni(NO_3)_2$ which is connected by a salt bridge to a solution of $Cu(NO_3)_2$ which contains a copper metal strip (the second electrode). An electric current flows when the electric circuit is completed by connecting a conductor to the two electrodes.

The anode (oxidation) half-reaction: $Ni(s) \rightarrow Ni^{2+} + 2e^-$
The cathode (reduction) half-reaction : $Cu^{2+} + 2e^- \rightarrow Cu(s)$
Net ionic equation: $Ni(s) + Cu^{2+} \rightarrow Ni^{2+} + Cu(s)$

17.47 The balanced equation is given in section 17.4. The C is oxidized from –2 to –1 and the chromium is reduced from +6 to +3.

$$8H^+ + 3C_2H_5OH + Cr_2O_7^{2-} \rightarrow 3C_2H_4O + 2Cr^{3+} + 7H_2O$$

$$0.0214\text{ L} \times \frac{0.100\text{ mol } Cr_2O_7^{2-}}{1\text{ L}} \times \frac{3\text{ mol } C_2H_5OH}{1\text{ mol } Cr_2O_7^{2-}} \times \frac{46.0\text{ g}}{1\text{ mol } C_2H_5OH} = 0.295\text{ g } C_2H_5OH$$

17.49 zero

17.51 a. In N_2O, the oxidation number of oxygen is -2 and there is 1 oxygen atom. Let the oxidation number of nitrogen be N.

$$2N + (-2) = 0$$
$$2N = +2$$
$$N = +1$$

The nitrogen has an oxidation number of +1.

b. In N_2O_3, the oxidation number of oxygen is -2 and there are 3 oxygen atoms. Let the oxidation number of nitrogen be N.

$$2N + 3(-2) = 0$$
$$2N - 6 = 0$$
$$N = +3$$

The nitrogen has an oxidation number of +3.

c. In N_2O_5, the oxidation number of oxygen is -2 and there are 5 oxygen atoms. Let the oxidation number of nitrogen be N.

$$2N + 5(-2) = 0$$
$$2N - 10 = 0$$
$$N = +5$$

The nitrogen has an oxidation number of +5.

d. The nitrogen in N_2, as an element, has an oxidation number of 0.

17.53 a. The oxidation number of Ca is +2 and O is –2.

$$+2 + S + 4(-2) = 0$$
$$S = +6$$

b. The oxidation number of O is –2 and the overall charge is –2.

$$S + 3(-2) = -2$$

$S = +4$

c. The oxidation number of O is –2 and the overall charge is 0.

$$S + 3(-2) = 0$$
$$S = +6$$

d. The oxidation number of O is –2 and the overall charge is 0.

$$S + 3(-2) = 0$$
$$S = +6$$

17.55 In TiO_2, the oxidation number of oxygen is -2 and there are 2 oxygen atoms. Let the oxidation number of titanium be Ti.

$$Ti + 2(-2) = 0$$
$$Ti - 4 = 0$$
$$Ti = +4$$

The titanium has an oxidation number of +4.

17.57 In $HClO_4$, the oxidation number of hydrogen is +1 and oxygen is -2; there are 4 oxygen atoms. Let the oxidation number of chlorine be Cl.

$$+1 + Cl + 4(-2) = 0$$
$$Cl = +7$$

The chlorine has an oxidation number of +7.

17.59 The hypochlorite ion, ClO^-, is acting as an oxidizing agent. The oxidizing agent is the one that gets reduced. Here, the oxidation number of Cl is reduced from +1 in ClO^- to -1 in the chloride ion, Cl^-.

17.61 H_2SO_4 is the oxidizing agent. (S atoms get reduced from +6 to -2 in H_2S.)
KI is the reducing agent. (I atoms get oxidized from -1 in KI to 0 in I_2.)

17.63 Step 1. Write half-reactions.

$$I^- \rightarrow I_2$$
$$HNO_2 \rightarrow NO$$

Step 2. Balance I atoms.

$$2\,I^- \rightarrow I_2$$
$$HNO_2 \rightarrow NO$$

Step 3. Balance O atoms by using 1 H_2O.

$$2\,I^- \rightarrow I_2$$
$$HNO_2 \rightarrow NO + H_2O$$

Step 4. Balance H atoms by using 1 H^+.

$$2\,I^- \rightarrow I_2$$
$$H^+ + HNO_2 \rightarrow NO + H_2O$$

Step 5. Balance electrical charges using 2 e^- and 1 e^-.

$$2\,I^- \rightarrow I_2 + 2e^-$$
$$1\,e^- + H^+ + HNO_2 \rightarrow H_2O + NO$$

Step 6. Electrons lost must equal electrons gained, and they do. Add the half-reaction equations and cancel electrons.

$$2\,I^- \rightarrow I_2 + 2e^-$$
$$\underline{2\,e^- + 2\,H^+ + 2\,HNO_2 \rightarrow 2\,H_2O + 2\,NO}$$
$$2\,I^- + 2\,H^+ + 2\,HNO_2 \rightarrow 2\,NO + 2\,H_2O + I_2$$

Step 7. Check to make sure all atoms and total charges on both sides are balanced and that all substances are in simplest integer ratios.

17.65 a. The half-reaction equation is: $Zn(s) \rightarrow Zn^{2+} + 2\,e^-$

b. Zinc is oxidized (to zinc ions) at this electrode.

c. Anode. (Oxidation takes place at the anode of an electrochemical cell.)

17.67 a. $Fe(s) \rightarrow Fe^{2+} + 2\ e^-$

b. oxidation, because electrons are lost

17.69 reduced

17.71 loss; gain

17.73 Step 1. Write half-reactions

$$V^{3+} \rightarrow VO_2^+$$
$$MnO_4^- \rightarrow Mn^{2+}$$

Step 2. V and Mn atoms are balanced.

Step 3. Balance O atoms by using 4 H_2O with Mn and 2 with V.

$$2\ H_2O + V^{3+} \rightarrow VO_2^+$$
$$MnO_4^- \rightarrow Mn^{2+} + 4\ H_2O$$

Step 4. Balance H atoms by using 8 H^+ with Mn and 4 with V.

$$2\ H_2O + V^{3+} \rightarrow VO_2^+ + 4\ H^+$$
$$8\ H^+ + MnO_4^- \rightarrow Mn^{2+} + 4\ H_2O$$

Step 5. Balance electrical charges using 2 e^- and 5 e^-.

$$2\ H_2O + V^{3+} \rightarrow VO_2^+ + 4\ H^+ + 2\ e^-$$
$$5\ e^- + 8\ H^+ + MnO_4^- \rightarrow Mn^{2+} + 4\ H_2O$$

Step 6. Electrons lost must equal electrons gained, so multiply the first equation by 5 and the second by 2. Add the half-reaction equations and cancel electrons (10 e^- on each side).

$$10\ H_2O + 5\ V^{3+} \rightarrow 5\ VO_2^+ + 20\ H^+ + 10\ e^-$$
$$\underline{10\ e^- + 16\ H^+ + 2\ MnO_4^- \rightarrow 2\ Mn^{2+} + 8\ H_2O}$$
$$5\ V^{3+} + 2\ H_2O + 2\ MnO_4^- \rightarrow 5\ VO_2^+ + 2\ Mn^{2+} + 4\ H^+$$

Step 7. Check to make sure all atoms and total charges on both sides are balanced and that all substances are in simplest integer ratios.

17.75 Step 1. Write half-reactions.

$$I^- \rightarrow I_2$$
$$Ce^{4+} \rightarrow Ce^{3+}$$

Step 2. Balance I atoms.

$$2\ I^- \rightarrow I_2$$
$$Ce^{4+} \rightarrow Ce^{3+}$$

Step 3. There are no oxygen atoms to balance.

Step 4. There are no hydrogen atoms to balance.

Step 5. Balance electrical charges using 2 e^- and 1 e^-.

$$2\ I^- \rightarrow I_2 + 2e^-$$
$$1\ e^- + Ce^{4+} \rightarrow Ce^{3+}$$

Step 6. Electrons lost must equal electrons gained, and they do. Add the half-reaction equations and cancel electrons.

$$2\ I^- \rightarrow I_2 + 2e^-$$
$$\underline{2\ e^- + 2\ Ce^{4+} \rightarrow 2\ Ce^{3+}}$$
$$2\ Ce^{4+} + 2\ I^- \rightarrow 2\ Ce^{3+} + I_2$$

Step 7. Check to make sure all atoms and total charges on both sides are balanced and that all substances are in simplest integer ratios.

17.77 a. I^-, it loses an electron

b. Cl_2, each Cl gains an electron

c. Cl_2, it gets reduced to Cl^- ions

d. NaI, I^- gets oxidized to I atoms

17.79 a. Zn, it loses 2 electrons

b. H^+, each H^+ gains an electron

c. HCl, each H^+ gets reduced to a H atom

d. Zn, it gets oxidized to Zn^{2+} ions

17.81 a. Fe, it loses 2 electrons

b. Cu^{2+}, each Cu^{2+} gains 2 electrons

c. $CuSO_4$, each Cu^{2+} gets reduced to a Cu atom

d. Fe, it gets oxidized to Fe^{2+} ions

CHAPTER 18
Fundamentals of Nuclear Chemistry

18.1 350; most elements have more than one naturally occurring isotope

18.3 ${}^{0}_{-1}e$; β

18.5 During alpha decay, the atomic number of the original nucleus decreases by two while the mass number decreases by four.

18.7 2; 4

18.9 Gamma rays—one type of high-energy radiation—have neither mass nor charge; they are emitted by unstable isotopes undergoing alpha or beta decay.

18.11 The superscript is the mass number; the subscript is the atomic number.

a. ${}^{131}_{53}I$ b. ${}^{3}_{1}H$ c. ${}^{1}_{0}n$ d. ${}^{222}_{86}Rn$

18.13 a. ${}^{0}_{-1}e$ b. ${}^{56}_{26}Fe$ c. ${}^{59}_{26}Fe$ d. ${}^{14}_{6}C$

18.15 Answers a and d represent isotopes (atoms with the same atomic number).

18.17 ${}^{214}_{84}Po \rightarrow {}^{4}_{2}He + {}^{210}_{82}Pb$

Lead is produced by this decay.

18.19 a. ${}^{238}_{92}U \rightarrow {}^{234}_{90}Th + {}^{4}_{2}He$

b. ${}^{234}_{90}Th \rightarrow {}^{234}_{91}Pa + {}^{0}_{-1}e$

c. ${}^{234}_{91}Pa \rightarrow {}^{234}_{92}U + {}^{0}_{-1}e$

d. ${}^{234}_{92}U \rightarrow {}^{230}_{90}Th + {}^{4}_{2}He$

18.21 Gamma rays will travel through a sheet of aluminum.

18.23 Alpha particles do not pass through rubber gloves.

18.25 The 4.5 s is one half-life, so ½ × 3000 atoms = 1500 atoms remain.
The 9.0 s is two half-lives, so ½ × ½ × 3000 atoms = 750 atoms remain.

18.27 Because 4/16 (or 1/4) of the activity remains, the sample has gone through 2 half-lives; the artifact's age is

$$2\ \text{half-lives} \times \frac{5730\ \text{y}}{\text{half-life}} = 11{,}460\ \text{y}.$$

18.29 When 1/4 of the activity remains, the sample will have gone through 2 half-lives, which will take

$$2\text{ half- lives x}\frac{13\text{ s}}{\text{half- life}}=26\text{ s}.$$

18.31 $\frac{1}{8}=\left(\frac{1}{2}\right)^3=3\text{ half - lives x}\frac{8.04\text{ days}}{1\text{ half - life}}=24\text{ days}$

18.33 $24\text{ hours x}\frac{1\text{ half} - \text{life}}{6\text{ hours}}=4\text{ half} - \text{lives}=\left(\frac{1}{2}\right)^4=\frac{1}{16}\text{x }176\text{ mg}=11.0\text{ mg}$

18.35 Ionizing radiation can cause vital molecules in living cells to be ionized or broken into fragments that cannot carry out normal functions.

18.37 Film badges.

18.39 Eighty

18.41 Radon-219 and radon-220 decay very quickly (their half-lives are measured in seconds), but radon-222, with a longer half-life of 3.82 days, is of greater concern; its daughters—alpha-emitting polonium radioisotopes—can be trapped in the lungs where they can damage cells and cause cancer.

18.43 4.5 x 10^9 yr; α particle; thorium-234

18.45 Lead-206 is formed from the alpha decay of polonium-210.

18.47 Radon, cosmic rays, rocks and soil, and radiation from inside our bodies.

18.49 3.8 days

18.51 ${}^{31}_{16}\text{S} \rightarrow {}^{0}_{+1}\text{e} + {}^{31}_{15}\text{P}$

18.53 a. ${}^{153}_{62}\text{Sm}$ b. ${}^{1}_{0}\text{n}$ c. ${}^{109}_{47}\text{Ag}$

18.55 a. ${}^{240}_{95}\text{Am}$ b. ${}^{242}_{96}\text{Cm}$ c. ${}^{243}_{97}\text{Bk}$ d. ${}^{245}_{98}\text{Cf}$

18.57 Small amounts of iodine-131 are administered for diagnostic purposes; the isotope is concentrated in the thyroid and the radiation emitted can be picked up by a detector and used to develop a photoscan of the tissue.

18.59 Cobalt-60 and radium-226 are radioisotopes used in cancer therapy.

18.61 No, the gamma radiation does not make food radioactive.

18.63 Americium-241 is used in certain smoke detectors (ionization type).

18.65 The critical mass is the minimum amount of a fissionable material required to sustain a chain reaction.

18.67 A moderator is used to slow down fast neutrons released by fissionable atoms; the slower-moving neutrons are captured by nuclei which, in turn, undergo fission. Heavy water and graphite have been used as moderators.

18.69 Neutrons; they are captured and released during chain reactions.

18.71 Uranium-235

18.73 When Fermi and Sergè bombarded uranium with neutrons, they thought they had produced a new element.

18.75 Differences between fission in nuclear bombs and fission in nuclear power plants: (1) Control rods are used in nuclear power plants, but not in bombs. (2) Nuclear bombs require a uranium-235 concentration of 85–90%, but nuclear power plants require uranium-235 enrichment to only 3%. (3) A moderator is used in nuclear power plants, but not in fission bombs.

18.77 During nuclear fusion, nuclei of small atoms must be brought together and united. Fusion takes place at the intense temperatures of the sun and in hydrogen bombs.

18.79 50 to 100 million °C

18.81 Characteristics of technetium-99^m that make it useful for diagnostic testing include its short half-life and its release of gamma rays-used to detect the presence of the isotope-with no accompanying alpha or beta particles that could cause unnecessary damage to the body.

18.83 a. Protons, 30; neutrons, 32; mass number, 62; symbol, Zn.

b. Protons, 94; neutrons, 147; mass number, 241 symbol, Pu.

c. Protons, 15; neutrons, 17; mass number, 32 symbol, P.

18.85 a. The missing particle is $^{11}_{5}B$ b. The missing particle is $^{12}_{6}C$

18.87 Three neutrons are released during this fission reaction. The reaction shows that one neutron is captured and three neutrons are released so a chain reaction is possible.

18.89 Important scientific events include Einstein's discovery that a small amount of matter could release a tremendous quantity of energy, Hahn and Strassmann's discovery that uranium had undergone fission in the experiment conducted by Fermi and Serge', Meitner's calculation that showed the great amount of energy released during fission of uranium, and Szilard's realization that neutrons released during fission could trigger the fission of other uranium atoms. These discoveries occurred during World War II, and Germany was working on the development of the bomb.

18.91 One neutron is released during this nuclear reaction.

18.93 PET is for positron emission tomography. A positron has a mass equal to that of an electron, but a positive charge. When a positron collides with an electron, 2 gamma rays are formed; they exit the body in opposite directions.

18.95 High energy radiation in the form of gamma rays can kill bacteria. In 1997 the U.S. Food and Drug Administration approved the use of irradiation using gamma rays on meat. The action follows a recall of a million pounds of frozen beef contaminated by E. coli bacteria.

18.97 Nuclear energy offers an alternative energy source to fossil fuels and has a variety of uses in medicine. With these advantages come the responsibilities of proper use and disposal.

18.99 Lise Meitner worked with Hahn on the discovery of nuclear fission. They also discovered element 91, protactinium.

CHAPTER 19
Organic Chemistry

19.1 With the development of thousands of synthetic organic compounds, the old definition of organic chemistry had to be expanded to include all carbon-containing compounds; originally the definition included only those compounds that are present in or produced by living organisms.

19.3 a. organic b. inorganic c. organic d. organic

19.5 a. NaOH, an inorganic compound, would be expected to have the higher melting point.

b. KCl, an inorganic compound, would be expected to have the higher melting point.

c. $C_{20}H_{42}$, with the higher molar mass, would be expected to have a higher melting point than C_6H_{12}.

19.7 Organic; because the substance is a liquid at room temperature, and because it burns, we know the substance is organic. Because the liquid was immiscible with water, we know it is a nonpolar molecular organic liquid.

19.9 a. 2; CH_3CH_3

b. 4; $CH_3CH_2CH_2CH_3$

c. 4; CH_3CHCH_3 (with CH_3 attached to the central C)

```
CH3CHCH3
    |
   CH3
```

d. 5; $CH_2{=}CHCH_2CH_2CH_3$

e. 8; $CH_3CH_2CH_2CH_2CCH_3$

```
                 CH3
                  |
CH3CH2CH2CH2CCH3
                  |
                 CH3
```

f. 2; $HC \equiv CH$

19.11 18

19.13 a. chloroform

b. hexane

c. bromoethane (ethyl bromide)

d. 2-chloropropane (isopropyl chloride)

19.15 a. same compound b. different compounds c. same compound

19.17 a. unsaturated

b. unsaturated

c. unsaturated

d. saturated

19.19 a. 3-methylpentane

```
CH3CH2CHCH2CH3
        |
       CH3
```

b. 2-methylpentane

```
CH3CHCH2CH2CH3
    |
   CH3
```

19.21 $CH_3CH_2CH_2CH_2OH$ primary

$CH_3CH_2CHCH_3$ (with OH on the third carbon) secondary

```
   CH3
   |
CH3C – OH      tertiary
   |
   CH3
```

```
CH3CHCH2 – OH   primary
   |
   CH3
```

19.23 butane

19.25

```
a. CH2=CHCHCH2CH2CH3
          |
       CH3CHCH3
```

```
b. CH3–C=C–CH3
        |  |
        Cl Cl
```

```
             Cl
             |
c. CH3–C=C–CH3
          |
          Cl
```

19.27

```
CH2CH2CH2CH3   1,2-dibromobutane
|   |
Br  Br
```

19.29 a. aromatic b. aliphatic c. aliphatic

19.31 Based on these structures, one would think benzene would behave like an alkene and undergo addition reactions, but this is not the case.

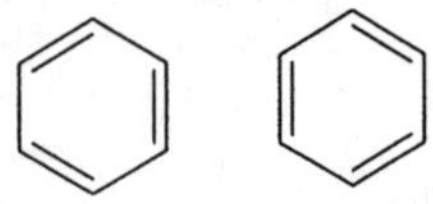

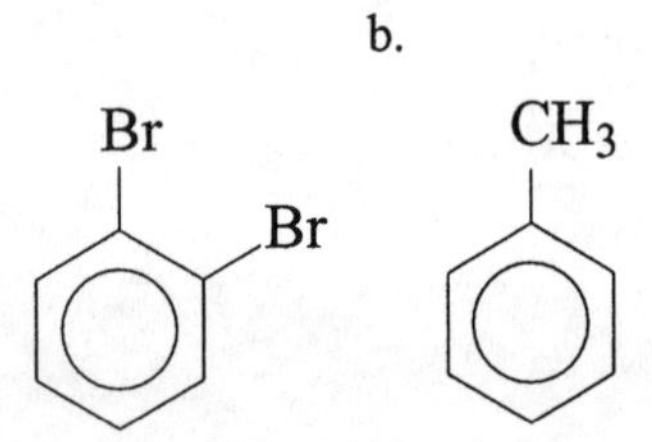

19.33 a. para b. ortho c. meta

19.35 a. b. c. d.

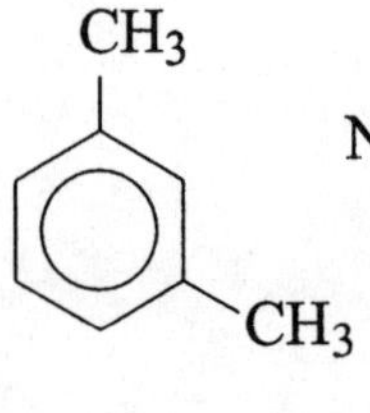

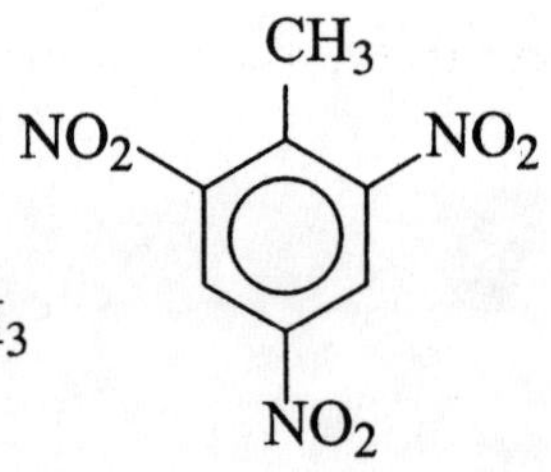

Br
Br

CH_3

CH_3
CH_3

CH_3
NO_2
NO_2
NO_2

19.37

```
a. CH3CHCH3   secondary alcohol
      |
      OH
```

```
b.    CH3  tertiary alcohol
      |
   CH3C–OH
      |
      CH3
```

```
c. CH3CHCH2–OH   primary alcohol
      |
      CH3
```

```
d. CH3CHCH2CH2–OH   primary alcohol
      |
      CH3
```

19.39 Destructive distillation involves heating a substance – such as wood – in the absence of oxygen. Methyl alcohol can be obtained from the destructive distillation of wood.

19.41 Denatured alcohol is ethanol that has been made unfit for use as a beverage by the addition of various poisonous and noxious substances.

19.43 40% (The percent alcohol by volume is one-half the beverage's proof.)

19.45 Commercial production of methanol involves the catalyzed reaction of hydrogen gas with carbon monoxide-obtained from partial oxidation of methane-at a high temperature and pressure.

$$CO + H_2 \xrightarrow{\text{catalyst-}} CH_3OH$$

19.47 1,2-Ethanediol (called ethylene glycol) is the most common diol. It is the key ingredient in automobile antifreeze, but it is also used extensively in the manufacture of polyester fibers and films.

19.49 The formula of phenol is C_6H_5OH. Phenol is used as an antiseptic and disinfectant.

19.51 R – O – R

19.53 ether; methyl *tert*-butyl ether

19.55 All aldehydes contain the carbonyl group.

```
  O
  ||
R–C–H
```

19.57

```
  O
  ||
H–C–H
```

Formaldehyde is used in embalming fluids, adhesives and plastics. Formalin is a 40% solution of formaldehyde in water.

19.59 a.

```
       O
       ||
CH3CH2CCH3
```

b.

```
   O
   ||
CH3CH
```

c.

```
    O CH3
    || |
CH3–C–CHCH3
```

19.61 A primary alcohol can be oxidized to give an aldehyde.

19.63 a.

```
CH3CHCH2CH2CH3
   |
   OH
```

b. CH_3CH_2OH

19.65

```
  O
  ||
R–C–OH
```

19.67

```
  O
  ||
R–C–O–R
```

19.69

```
             O                                    O
             ||                                   ||
CH3CH2CH2C–OH  +  HO–CH3  →  CH3CH2CH2–C–OCH3  +  H2O
butyric acid      methyl alcohol   methyl butyrate      water
```

19.71 polyetylene terephthalate

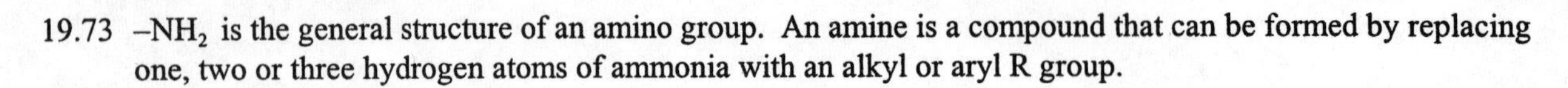

19.73 $-NH_2$ is the general structure of an amino group. An amine is a compound that can be formed by replacing one, two or three hydrogen atoms of ammonia with an alkyl or aryl R group.

19.75 a. $CH_3CH_2NH_2$ primary b. CH_3NCH_3 (with H on N) secondary c. CH_3-N-CH_3 (with CH_3 on N) tertiary

```
                    |                    |
                    H                   CH3
```

19.77 This is an amide.

```
     O
     ||
CH3C–NHCH3
```

19.79 primary amine

19.81 Charcoal lighter fluid; it is a nonpolar organic petroleum product that will, therefore, dissolve other nonpolar petroleum products such as motor oil and heavy greases.

19.83 Methanol is metabolized to formaldehyde an its long term toxicity is quite high; furthermore, ingestion of as little as 30 mL of methanol can cause permanent blindness so it cannot be used as a solvent for medicines. Ethanol can be metabolized – within limits-without these problems.

19.85 a. 2-butanol b. 2-pentanol c. 3-methyl-1-butanol d. 2-ethoxyethane (diethyl ether)

19.87 a. ethyl methyl ketone b. propanal c. 3-hexanone

19.89 a. butanoic acid

```
          O
          ||
CH3CH2CH2C–OH
```

b. dodecanoic acid

```
                                          O
                                          ||
CH3CH2CH2CH2CH2CH2CH2CH2CH2CH2CH2C–OH
```

c. ethyl butanoate

```
          O
          ||
CH3CH2CH2COCH2CH3
```

19.91 a. propylamine; a primary amine

b. ethylmethylamine; a secondary amine

c. trimethylamine; a tertiary amine

19.93 2,2,4-trimethylpentane (isooctane), which burns smoothly, is given an octane rating of 100. Heptane, which burns with knocking, is given an octane number of 0. The gasoline is given an octane rating based on comparison to a rating of a mixture of isooctane and octane.

19.95 30 °C to 200 °C, C_5 to C_{12}; straight run gasoline is obtained directly from distillation.

19.97 Polymers are very large molecules made by linking together many smaller molecules.

19.99 PVC (polyvinylchloride) is made from vinyl chloride and is used in plastic PVC pipes, plastic wrap, and garden hoses.

19.101 Polyethylene is made by linking ethylene molecules end to end. PETE, polyethylene terephthalate, is prepared by alternately linking ethylene glycol with terephthalic acid.

19.103 –OH, R–OH

19.105

```
 O          O
 ||         ||
–C–H,     R–C–H
```

19.107 R–O–R

19.109 A primary amine has one alkyl group attached to the nitrogen. A secondary amine has two alkyl groups attached to the nitrogen atom.

19.111 a. ether b. ketone c. alkyne d. ester

CHAPTER 20
Biochemistry

20.1 Dextrose or "blood sugar"; it is a monosaccharide.

20.3 The orientation of –OH and –H at C #4.

20.5 a. fructose

b. sucrose (made of a fructose and glucose unit)

20.7 alpha, beta and open-chain

20.9 Amylose, amylopectin, and glycogen are polysaccharides made up of glucose units joined by alpha linkages.

20.11 In starch, the glucose units are joined by alpha linkages, but in cellulose, the glucose units are joined by beta linkages.

20.13 The starch present in the soda cracker is broken up to give glucose units even while the cracker is in the mouth, and the glucose tastes sweet.

20.15 The configuration shown is L.

20.17 A saturated fatty acid is a long-chain carboxylic acid with its carbon atoms joined by single bonds. Examples: stearic and oleic acids.

20.19 omega-6

20.21 A simple lipid is composed of a glycerol molecule joined to three fatty acid molecules by ester linkages.

20.23 a. monounsaturated; olive oil

b. saturated; palm oil

c. saturated; oil of nutmeg

d. polyunsaturated; liver oil

20.25 The iodine number is an overall measure of unsaturation of fats and oils.

20.27 The iodine number for a liquid margarine would be expected to be higher than the iodine number for a solid margarine. Liquids (oils) are more unsaturated than are solids (fats).

20.29 Amino acids are the simplest building units of proteins; the amino acids are joined by peptide bonds.

20.31 An amino acid contains a carboxyl group, –COOH, and an amino group, $-NH_2$.

20.33 About 20 different amino acids are incorporated into proteins.

20.35 A polypeptide is made up of a string of about 10 or more amino acids. When the molar mass (molecular weight) of the polypeptide exceeds 10,000, it is called a protein.

20.37 The sequence of amino acids in a protein is the primary structure.

20.39 Deoxyribonucleic acid (DNA) and ribonucleic acid (RNA) are the two types of nucleic acids. The DNA is found in the cell nucleus; the RNA is found in all parts of the cell.

20.41 Ribose and deoxyribose are the two sugars that can be in a nucleotide.

20.43 DNA is composed of a double-stranded helix. The two helixes are wound about one another and are held together by strong hydrogen bonding between the paired bases.

20.45 A vitamin is a specific organic compound that is required in the diet in small amounts for proper body functioning; the shortage of a vitamin results in a vitamin-deficiency disease.

20.47 Vitamins A, D, E, and K are the fat-soluble vitamins; they are nonpolar.

20.49 a. Prolactin is a protein that maintains the production of estrogens and progesterone; it stimulates the formation of milk.

b. Insulin is a protein that increases cell usage of glucose and increases glycogen storage.

c. Aldosterone is a steroid that regulates salt metabolism and stimulates the kidneys to retain Na^+ and excrete K^+.

d. Progesterone is a steroid that regulates the menstrual cycle and maintains pregnancy.

20.51 Glucose is the monosaccharide obtained from the hydrolysis of maltose.

20.53 a. sucrose b. glucose

20.55 Chemists and biochemists help in fighting genetic disease when they work to find which genes are linked to specific diseases. For example, a U.S Canadian research team identified the gene that causes cystic fibrosis, a common lethal genetic disease.

20.57 Ascorbic acid is vitamin C, calciferol is vitamin D_2, retinol is vitamin A, and tocopherol is vitamin E.

20.59 A chiral carbon has four different groups attached. Glucose has four chiral carbons. They are carbons 2, 3, 4, and 5.

20.61 Triglycerides, or simple lipids, are esters derived from glycerol and long chain carboxylic acids (fatty acids).

20.63 The *trans* configuration is a straight chain molecule compared to the *cis* configuration which is a bent molecule.

20.65 HDL (high-density lipoprotein) cholesterol is thought of as the "healthy" cholesterol and LDL (low-density lipoprotein) cholesterol is the bad cholesterol.

20.67 Alpha acetal linkage

20.69 Mutarotation

20.71 a. both fat and oil b. fat c. oil d. fat

20.73 No. A fatty acid is a carboxylic acid – saturated or unsaturated – that combines with glycerol by ester linkages to form triglycerides called fats. These fats are also called simple lipids.

20.75 a. polyunsaturated, omega-3 b. monounsaturated

20.77 a. beef fat b. soybean oil c. liver

20.79 a. vitamin A b. amino acid c. steroid d. nucleic acid component

20.81 a. nucleic acid component b. amino acid c. polysaccharide d. disaccharide